OPC UA

The Everyman's Guide to the Most Important Communications Architecture of Industrial Automation

JOHN RINALDI

JOHN RINALDI

Copyright © 2016 JOHN RINALDI

All rights reserved.

ISBN-13: 978-1530505111
ISBN-10: 1530505119

DEDICATION

To the Automation Engineer, the unsung hero of American Manufacturing.

JOHN RINALDI

THE EVERYMAN'S GUIDE TO OPC UA

TABLE OF CONTENTS

DEDICATION..iii
TABLE OF CONTENTS...v
TABLE OF FIGURES..vii
TABLE OF TABLES..viii
ACKNOWLEDGMENTS..i
FOREWORD..2
INTRODUCTION..4
A LITTLE OPC UA HISTORY.....................................10
FACTORY VS IT AUTOMATION...............................16
10 THINGS TO KNOW ABOUT OPC UA..................20
THE UA SERVER...27
THE UA CLIENT..36
INTRODUCTION TO SECTION II..............................42
TRANSPORT LAYERS..43
DEVICE DISCOVERY...49
ADDRESS SPACE..64
INFORMATION MODELING......................................76
DATA TYPING...88
DATA ENCODINGS...100

SECURITY	103
OPC UA SERVICES	121
OPC UA IN PRACTICE	127
THREE USE CASES	133
COMPETITIVE TECHNOLOGIES	140
TERMS TO KNOW	151
RESOURCES	158
ABOUT THE AUTHOR	159

LEARN MORE Have a little fun and get some relevant information. Sign up for the Real Time Automation Newsletter

http://www.rtaautomation.com/company/newsletter/

THE EVERYMAN'S GUIDE TO OPC UA

TABLE OF FIGURES

Figure 1 - OPC Classic Architecture 12
Figure 2 - OPC UA Server Overview 27
Figure 3 - OPC UA Address Space 29
Figure 4 - OPC UA Server Software Architecture 31
Figure 5 - Server Object in an OPC UA device 32
Figure 6 - Client Server Relationships 37
Figure 7 OPC UA Server with three endpoints 53
Figure 8 - Discovery Process with No LDS Server 58
Figure 9 - Discovery Process with LDS Server 60
Figure 10 - Discovery Process with LDS-ME Server 62
Figure 11 - The Structure of an OPC UA Node 64
Figure 12 - Object & Variable node Class example 66
Figure 13 - Heater Object Address Space 73
Figure 14- Address Space creation process 74
Figure 15 - A snippet of a pump Information Model 77
Figure 16 - The Wikipedia Information Model 79
Figure 17 - Curing oven Information Model 80
Figure 18 - Curing oven Info Model with Standard Nodes 85
Figure 19 - Service Request Mapping Client to Server 107
Figure 20 - Endpoints and Security 115
Figure 21 - Secure Conversation / TCP Transport (identical structure) ... 116
Figure 22 - The Secure Conversation Message Body 117
Figure 23 - OPC Secure Conversation and Security 118
Figure 24 - Create Session Security 119
Figure 25 - Activate Session Security Operations 120
Figure 26 – Root Folder and its Object folders 124
Figure 27 - Motor Controller Connected to Enterprise 134

Figure 28 - Traditional factory data to the enterprise 135

TABLE OF TABLES

Table 1 - Well Known Discovery endpoints 54
Table 2 - Get Endpoints Description 55
Table 3 - Register Server Service Overview 56
Table 4 - Find Servers Service Overview 57
Table 5 - Node Class table 70
Table 6 - Mandatory and Optional Attributes 71
Table 7 - OPC UA Common Data Types 91
Table 8 - *NodeID* Type Structure 93
Table 9 - NodeID Derived Type 94
Table 10 – Extension Object Derived Type 95
Table 11 – Request Header Derived Type 95
Table 12 – Response Header Derived Type 96
Table 13 – Application Description Derived Type 97
Table 14 – Endpoint Description Derived Type 98
Table 15 - LocalID Derived Type 98
Table 16 - ReadValueID Derived Type 99
Table 17 - Digital Application Interface Certificate 109
Table 18 – Digital Software Certificate 110
Table 19 - Message Security Mode 110
Table 20 - Message Security Policy 111
Table 21 - User Token Policy 112
Table 22 - Session ID 113
Table 23 - User Identity Token 113
Table 24 - Channel Security Token 114
Table 25 - Session Authentication token 114
Table 26 - OPC UA Service Sets 122
Table 27 - Example In line Data Definition 126

THE EVERYMAN'S GUIDE TO OPC UA

ACKNOWLEDGMENTS

This book would not be possible without the dedication, friendship, persistence, support and follow through of the entire staff at Real Time Automation. Special thanks to Drew Baryenbruch for the cover designs of all my books and freeing me of daily sales and marketing so that I can take on projects like this.

By reading and accepting this information you agree to all of the following: You understand that this is simply a set of opinions (and not advice). This is to be used for entertainment, and not considered as "professional" advice. You are responsible for any use of this information in this work and hold the author and all members and affiliates harmless in any claim or event.

FOREWORD

When we first started OPC in 1995, the focus was to provide a simple open standard to allow the first tier visualization applications to seamlessly be able to read/write and subscribe to data from factory automation and process control devices. 20 years of constant technology change and persistent opportunities for interoperability in Industrial Automation demands plug-and-play products for interoperable solutions even more than ever!

The initial focus of OPC UA was to address interoperability in Industrial Automation and related domains. Suppliers, end users, and other organizations (some proficient in Industrial Automation and some not) have now recognized the value proposition and have successfully leveraged the OPC UA technology beyond just Industrial Automation. The OPC Foundation has provided a solid solution for all types of automation, and John's ability to look back and also predict the future has resulted in him rolling out another edition of a fountain of knowledge.

The vision of OPC UA now is all about addressing the Industrial Internet of Things, providing a complete solution for moving data and information from the embedded world to the enterprise. OPC UA is well-positioned to address all the needs of Industrie 4.0, the Industrial Internet of Things, the Internet of Things, M2M and B2B. It's all about information integration from the embedded world to the cloud. The OPC Foundation has been actively collaborating with many organizations to facilitate modeling of their data to provide a unified mechanism for transporting their complex Information Models seamlessly between disparate systems. As of the writing of this book the OPC Foundation has already been actively engaged with numerous organizations inclusive of petrochemical, pharmaceutical, security, Building Automation, machine builders, and wastewater treatment.

I've had the opportunity to review this book before publication, and I'm delighted to say that John has done a superb job in providing an excellent overview of all of the OPC UA technology, and more importantly a solid foundation for both suppliers and end users to truly understand how to maximize their use of the OPC UA technology. The book provides an excellent foundation for readers that are either developing or using systems based on the OPC UA technology. The reader of this book is fortunate enough to have an opportunity to understand the philosophy of John Rinaldi through his creative writing style. This book is one of those books where you can sit down and read it cover to cover and then use it as your reference material from the past to the present to the future!

The architecture and design of OPC UA provides complete support for OPC products that were developed in 1995, and will continue to provide the solid infrastructure to allow products to continue to be developed essentially having timeless durability from now and into the future even as technology continues to change on a daily basis.

Consumer electronics and the engineers of tomorrow are really driving the future of automation, and OPC UA is well-positioned to provide highly integrated systems without requiring complex software development and constant reengineering. The future is here today for data and information integration between any connectable device.

The Internet of Things is about connectivity and adding OPC UA to it provides a mechanism for intelligent connectivity through information integration.

The OPC Foundation is a nonprofit international standards organization truly dedicated to providing the best specifications, technology, process and certification. Having excellent writers like John to tell the story and help you maximize use of the technology is of the utmost importance.

All of us in the OPC Foundation community thank you for choosing this book and taking time to read it to understand how to best use the OPC UA technology and truly to be able to have best-of-breed products and solutions for information integration and interoperability.

Thanks again John for writing such a great book!

Thomas J Burke
OPC Foundation President & Executive Director
March, 2016

INTRODUCTION

What is OPC UA is a very simple question. The answer when you are discussing a complex technology architecture like OPC UA isn't as simple.

OPC UA which I will refer to as UA throughout this book is the next generation of OPC technology. UA is a more secure, open, and reliable mechanism for transferring information between Servers and Clients. It provides more open transports, better security, and a more complete Information Model than OPC, which I will refer to as OPC Classic. UA provides a very flexible and adaptable mechanism for moving data between enterprise type systems and the kinds of controls, monitoring devices, and sensors that interact with real world data.

Why a totally new communication architecture? OPC Classic is limited and not well-suited for today's requirements to move data between enterprise/Internet systems and the systems that control real processes that generate and monitor live data. These limitations include:

- Platform dependence on Microsoft – OPC Classic is built around DCOM (Distributed Component Object Model) an older communication technology that is being de-emphasized by Microsoft.
- Insufficient Data Models – OPC Classic lacks the ability to adequately represent the kinds of data, information, and relationships between data items and systems that are important in today's connected world.
- Inadequate Security – Microsoft and DCOM are perceived by many users to lack the kind of security needed in a connected world with sophisticated threats from viruses and malware.

> **LEARN MORE** You can get a link to YouTube Videos introducing OPC UA on the resources web page for this book. Just visit:
>
> http://www.rtaautomation.com/technologies/opcuainfo/

UA is the first communication technology built specifically to live in that "No Man's Land" where data must traverse firewalls, specialized platforms, and security barriers to arrive at a place where that data can be turned into information. UA is designed to connect databases, Analytic tools, Enterprise Resource Planning (ERP) systems and other enterprise systems with real world data from low end controllers, sensors, actuators, and monitoring devices that interact with real processes that control and generate real world data.

UA uses scalable platforms, multiple security models, multiple transport layers, and a sophisticated Information Model to allow the smallest dedicated controller to freely interact with complex, high end Server applications. UA can communicate anything from simple downtime status to massive amounts of highly complex plant wide information.

UA is a sophisticated, scalable, and flexible mechanism for establishing secure connections between Client and Servers. Features of this unique technology include:

Scalability – UA is scalable and platform independent. It can be supported on high end Servers and on low end sensors. UA uses discoverable profiles to include tiny embedded platforms as Servers in a UA system.

A Flexible Address Space – The UA Address Space is organized around the concept of an Object. Objects are entities that consist of Variables and Methods and provide a standard way for Servers to transfer information to Clients.

Common Transports and Encodings – UA uses standard transports and encodings to ensure that connectivity can be easily achieved in both the embedded and enterprise environments.

Security – UA implements a sophisticated Security Model that ensures the authentication of Client and Servers, the authentication of users and the integrity of their communication.

Internet Capability – UA is fully capable of moving data over the Internet.

A **Robust Service Set** – UA provides a full suite of services for Eventing, Alarming, Reading, Writing, Discovery and more.

Certified Interoperability – UA certifies profiles such that connectivity between a Client and Server using a defined profile can be guaranteed.

A sophisticated Information Model – UA profiles more than just an Object Model. UA is designed to connect Objects in such a way that true Information can be shared between Clients and Servers.

Sophisticated Alarming and Event Management – UA provides a highly configurable mechanism for providing alarms and event notifications to interested Clients. The Alarming and Event mechanisms go well beyond the standard change-in-value type alarming found in most protocols.

Integration with Standard Industry-specific Data Models – The OPC Foundation is working with a number of industry trade groups that define specific Information Models for their industries to support those Information Models within UA.

How OPC UA Differs From Plant Floor Systems

I've studied this technology for a long time now. And yet there is a question that I almost shirk from. In fact, I sometimes hate to answer it.

It's not because I don't understand what OPC UA is. It's not that I don't understand how it works. And it's not that I don't believe that it is a very valuable tool to almost every plant floor system.

It's just hard to put it into context when there isn't anything to compare it to. For example, when ProfiNet IO came out, I could tell people that it's the equivalent of EtherNet/IP for Siemens controllers. Same kind of technology. Basically the same kind of functionality. Easy to explain.

But how do I explain UA when it doesn't have an equivalent? You could say that it's Web Services for automation systems. Or that it's SOA for automation systems, an even more arcane term. SOA is Service Oriented Architecture, basically the same thing as Web Services. That's fine if you're an IT guy (or gal) and you understand those terms. You have some context.

But if you're a plant floor guy, it's likely that even though you use Web Services (it's the plumbing for the Internet) you don't know what that term means.

So the reason I get skittish about answering this question is that they always follow up with another question that makes me cringe: "Why do we need another protocol? Modbus TCP, EtherNet/IP and ProfiNet IO work just fine."

So I have to start with the fact that it's not like EtherNet/IP, ProfiNet IO, or Modbus TCP. It's a completely new paradigm for plant floor communications. It's like trying to explain EtherNet/IP to a PLC programmer in 1982. With nothing to compare it to, it's impossible to understand.

That's where I am trying to explain OPC UA.

The people I'm trying to reach have lived with the PLC networking paradigm for so long that it's second nature. You have a PLC; it is a Master kind of device and it moves data in and out of Slave devices. It uses really simple, transaction-type messaging or some kind of connected messaging.

In either case, there is this buffer of output data in a thing called a Programmable Controller. There is a buffer of input data in a bunch of devices called Servers, Slaves or nodes. The buffer of input data moves to the Programmable Controller. The output data buffers move from the Programmable Controller to the devices. Repeat. Forever. Done.

That's really easy to wrap your mind around. Really easy to see how it fits into your manufacturing environment and really easy to architect.

OPC UA lives outside that paradigm. Well, really that's not true. OPC UA lives in parallel with that paradigm. It doesn't replace it. It extends it. Adds on to it. Brings it new functionality and creates new use cases and drives new applications. In the end it increases productivity, enhances quality, and lowers costs by providing not only more data but information and the right kind of information to the production, maintenance, and IT systems that need that information when they need it.

Pretty powerful, huh?

Our current mechanisms for moving plant floor data – few or no systems move information – is brittle. It takes massive amounts of human and computing resources to get anything done. And in the process we lose lots of important meta-data, we lose resolution, and we create fragile systems that are nightmares to support.

And don't even ask about the security holes they create. Because when there are problems (and there always are) the first thing everyone does is to remove the security and reboot.

These systems are a fragile house of cards. They need to be knocked down.

And because of all this, opportunities to mine the factory floor for quality data, interrogate and build databases of maintenance data, feed dashboard reporting systems, gather historical data, and feed enterprise analytic systems are lost. Opportunities to improve maintenance procedures, reduce downtime, compare performance at various plants, lines, and cells across the enterprise are all lost.

This is the gap that OPC UA fills. It's not something ProfiNet IO can do even though the devoted acolytes would contest that statement. It's not something that EtherNet/IP can do. And it'd be a joke to talk about Modbus TCP in this context.

So I'm back to the original question, "What exactly is OPC UA"?

OPC UA is about reliably, securely and most of all, easily, modeling "Objects" and making those Objects available around the plant floor, to enterprise applications and throughout the corporation. The idea behind it is infinitely broader than anything most of us have ever thought about before.

And it all starts with an Object. An Object that could be as simple as a single piece of data or as sophisticated as a process, a system or an entire plant.

It might be a combination of data values, meta-data, and relationships. Take a Dual Loop Controller. The Dual Loop Controller Object would relate variables for the setpoints and actual values for each loop. Those variables would reference other variables that contain meta-data like the temperature units, high and low setpoints and text descriptions. The Object might also make available subscriptions to get notifications on changes to the data values or the meta-data for that data value. A Client accessing that one Object can get as little data as it wants (single data value) or an extremely rich set of information that describes that controller and its operation in great detail.

OPC UA is, like its factory floor cousins, composed of a Client and a Server. The Client device requests information. The Server device provides it. But as we see from the Loop Controller example, what the UA Server does is much more sophisticated than what an EtherNet/IP, Modbus TCP or ProfiNet IO Server does.

An OPC UA Server models data, information, processes, and systems as Objects and presents those Objects to Clients in ways that are useful to vastly different types of Client applications. And better yet, the UA Server provides sophisticated services that the Client can use including:

- Discovery Services – services that Clients can use to know what Objects are available, how they are linked to other Objects, what kind of data and what type is available, what meta-data is available that can be used to organize, classify, and describe those Objects and Values.
- Subscription **Services** – services that the Clients can use to identify what kind of data is available for notifications. Services that Clients can use to decide how little, how much and when they wish to be notified about changes (not only to data values but to the meta-data and structure of Objects).
- Query **Services** – services that deliver bulk data to a Client, like historical data for a data value.
- Node **Services** – services that Clients can use to create, delete, and modify the structure of the data maintained by the Server.
- Method **Services** – services that the Clients can use to make function calls associated with Objects.

Unlike the standard industrial protocols, an OPC UA Server is a data engine that gathers information and presents it in ways that are useful to various types of OPC UA Client devices. Those devices could be located on the factory floor like an HMI, a proprietary control program like a recipe manager or a database, dashboard, or a sophisticated analytics program that might be located on an enterprise Sever.

Even more interesting, this data is not necessarily limited to a single physical node. Objects can reference other Objects, Data Variables, Data Types, and more that exist in nodes off someplace else in the subnet or someplace else in the architecture or even someplace else in the Internet.

OPC UA *Organizes* processes, systems, data and information in a way that is absolutely unique to the experience of the Industrial Automation industry. It is a unique tool that attacks a completely different problem than that solved by the EtherNet/IP, Modbus TCP and Profinet IO Ethernet protocols. UA is an Information Modeling and delivery tool that provides access to that information to Clients throughout a plant, the enterprise or on the Internet.

A LITTLE OPC UA HISTORY

If you've been around a while, you probably can remember when Bill Clinton was this delusional governor of the tiny state of Arkansas who thought that somehow he could become a president of the United States. If that's you, then you probably also remember when PCs started to show up on the factory floor.

The PC on the factory floor was a big deal. A really big deal. People were touting that PCs would replace Programmable Controllers. Easy connectivity. Standard programming languages. Cheap hardware.

A number of companies were created to exploit that technology. Some, like Event Technologies, with their GELLO software, failed. Others like Universal Automation's FloPro, Think & Do and Steeplechase were absorbed into bigger companies and eventually faded away. Others like Wonderware not only thrived, but were massively successful.

Remember, this was the 1990s. Very early in the development of the personal computer. Actually very early in the development of Windows. We're talking Windows 3.0, Windows 95, and Windows for Work Groups.

This stuff was, to say the least, challenging to work with. Nothing integrated. Every application was a standalone application. Stuff crashed. Hard drives didn't last long. But that was okay, because every nine months or so there was a new something to buy. New drives, new processor, new Windows.

It was a time of unprecedented change. Impossible for anyone not working full time in the PC industry to keep track of.

In the midst of all that, people, usually systems integrators, were trying to incorporate these computers into factory floor applications. Looking back, it's kind of funny how they stood on their heads to integrate them. Special keyboard covers, cabinets, fans and filters. These were delicate machines, after all, with a lot of mechanical pieces.

And these brave souls even had to deal with labor issues. I recall that at some Procter & Gamble plants, unions demanded wage adders for anyone having to touch a computer keyboard. Increased salary came with a job where you worked with a computer.

They did all this to avoid what they considered overpriced Programmable Controllers. They figured they could write specialized applications to, if not replace those PLCs, and bring a more sophisticated level of logic to these applications.

One of the things they incorporated was a lot of serial communications (long before Ethernet). PLCs were never very good with serial; not enough serial ports and lousy logic instructions to process the serial data. Any application that used a serial device was a candidate for a PC instead of a PLC.

To implement these applications, the programmers had to write drivers for the serial devices. Need to get a flow rate out of a scale or a flow meter? Or send the special control codes to a device and get a response back? Lots of programming for some of these devices. The messaging usually wasn't well conceived or concise or easy to troubleshoot.

These drivers could be pretty tricky. Years and years ago, I wrote serial drivers for lots of loop controllers. You had to send control codes to select a specific Loop Controller from the network (RS485 usually). Then you waited for a response. If a response didn't come, you had to take some action to let the application know you had processed the data.

Lots of times, you set up loops to run through a bunch of these. Sometimes, throughput or response time was crucial. Remember, we were doing this on old, slow PCs in the 1990s. A programmer could spend lots of time at this. Then there was troubleshooting in the field. Trying to figure out why only fourteen out of fifteen devices worked. Nightmare time.

And if you were a systems integrator, you were doing this over and over for every new project! Not a lot of those project managers slept well in those days.

But help was on the way. Around 1994-1995, a bunch of factory automation experts got together and decided there had to be a better way. Their work eventually led to the OPC 1.0 and the creation of the OPC Foundation.

Their mission was really pretty simple: create a way for applications to get at data inside an automation device without having to know anything about how that device works.

Pretty challenging mission. Nothing like this had ever been done before. How to make the thousands of automation devices accessible to any PC application?

Their solution centered on COM (Component Object Model). COM was a fairly new Microsoft technology at the time. It provided a way for

Microsoft Windows applications to share data. One application could request data, and another would supply it.

The OPC founding fathers took this technology mated it to an API that supported device protocols for automation devices, and OPC 1.0 was born. Figure 1 illustrates what this looked like when done.

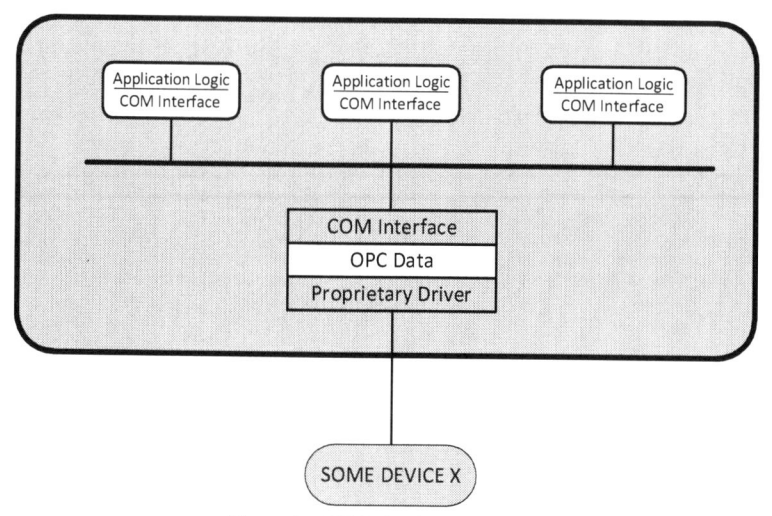

Figure 1 - OPC classic architecture

Data from the device, no matter what physical layer or protocol it supported, could be read or written by any Microsoft application that supported COM.

This was revolutionary and probably accelerated the adoption of PCs on the factory floor. For the first time ever, systems integrators and other factory floor application developers could implement applications without spending vast sums on driver development. Those project managers suddenly started sleeping a lot better.

And this technology made some people rich. Not Mark Zuckerberg rich, but "tollgate" position rich. Companies like Kepware, Matrikon and others owned the tollgate between your Microsoft application and all those automation devices. They began creating vast numbers of OPC Servers that could be incorporated into all sorts of applications. The more OPC Servers they created, the more applications people built.

But all wasn't well in OPC land. Soon, the applications demands and a changing automation landscape began to erode the acceptability of PCs and OPC Servers in these applications.

Why?

People mostly, and COM (DCOM is the distributed variant), the Microsoft technology underpinning OPC development. Let me explain.

COM is difficult to maintain and understand without significant training.

And how you use it, configure it and set up authorizations varies slightly from one version of Microsoft Windows to the next. There are many ways to screw it up. And when you screw it up, your OPC Server stops transferring data, and the people who want the data start screaming.

It's actually more insidious than that. If a well-meaning but undertrained individual (not usually understanding that they are undertrained) goes in and fiddles with some COM or DCOM (Distributed version of COM) parameters, nothing happens. Literally, there is no effect. The OPC Server keeps working (for a while anyway).

Eventually, a week, a month or three months later, there's a reboot. Maintenance powered down the manufacturing cell, a year-end shutdown, whatever, but the machine is powered down. Now, when you bring the machine back up, all of a sudden, for "no reason at all," DCOM stops working. The memory of that guy fiddling with the DCOM parameters is long forgotten. Instead, the OPC Server is blamed. It just stopped working.

It's a nightmare. The system needs to be fixed and another well-meaning, undertrained person starts fooling around. First thing they do is remove all the security. Generally, that makes it work. "Whew," they say. "Glad I could get that fixed."

Now you're set up for real trouble. There's a security hole in one of your Servers that can lead a hacker down a path to who knows where.

And that's the kind of thing that makes the VP of IT lay in bed at night staring at the ceiling. He's the one whose butt is going to be called on the carpet if that hacker ever does attack.

And it does happen. Someone without malicious intent finds a "blank" USB stick and eventually plugs it into this now-unprotected Server. Malware then starts looking for paths to specific automation equipment, and all of a sudden you have a much bigger problem on your hands.

But in truth, the lack of DCOM knowledge and the seemingly inconsequential act of plugging in a data stick are really not OPC Classic problems. They're management issues. Management didn't dictate that a checklist be in place when an OPC Server stops communicating. Management didn't have a certification program in place ensure that the people maintaining OPC Servers were well trained in COM and DCOM and troubleshooting OPC Server problems.

It's just more convenient to blame the technology than ourselves and our management. So OPC has taken some hits over the past few years in its public perception.

Probably not warranted, but what do they always say? "Perception is reality."

But there's another more pervasive problem with OPC. One that can't be blamed on management. It's the deficiencies that come with dependency on Microsoft and Windows.

COM, the base technology for OPC, is a Microsoft product. It runs only on Microsoft platforms. Not Linux, not VxWorks, not anything else. And that's a problem.

Microsoft has a well-deserved negative reputation, especially in Industrial Automation. In this industry, we generally build automation processes to last. There are a few products that are short-lived, but it's much more common to build production processes for diapers, soap, tea, and hundreds of other products that we're going to run for the next five, ten or twenty years.

Microsoft products and PCs aren't suited for that kind of environment. Every time you buy a new laptop, you are hopelessly obsolete in, what, six months? How do you maintain an OPC Server on a Microsoft Windows platform for the next ten years?

So there's been a desire to run OPC Servers on other platforms. Platforms with longer lives and stable hardware that will last. Platforms that are smaller. Embedded platforms. The question being asked was "Why do I have to be tied to Microsoft? What about Linux? Why can't my flow meter and lots of other devices be an OPC Server?"

None of that is possible with Microsoft and DCOM.

There's also been dissatisfaction with OPC in the way that data gets to all those data-hungry Servers in the upper echelons of the factory automation system and at the enterprise.

Most data is passed to those systems through a PLC. Possibly it starts in an RFID reader passing a pallet number, ID code and weight from the RFID reader to the PLC. From the PLC, it gets read by an OPC Server and passed to another application in the PC that transfers it to a logging database in the enterprise.

The real problem with the original version of OPC (what is now sometimes called OPC Classic) is that this is an expensive and inefficient way to get data from a device (RFID reader) into that database. There's a PC involved – someone has to set it up, maintain it, validate that it is secure, etc. There is the hardware setup and labor and the ongoing labor to maintain it.

But more than that, it's very inefficient and provides incomplete data. The data has to be carefully managed all the way from the RFID reader to the Server to make sure that the different systems use the correct data types, that resolution is maintained, the endian order (which byte is first) is proper for that system. It's not easy.

And every time you decide you want a new piece of data, you have to touch multiple systems without breaking any of them. "Yuck" is the technical term for it.

Plus, you don't get any of the meta-data. Meta-data being the associated data that provides the semantics for what you are transferring. Meta-data includes stuff like units, scaling and all that other stuff that lets you work with the data without guessing as to what it is.

So, even though OPC Classic is wildly successful and works well when managed right, there was enough dissatisfaction with the security issues, platform issues and data inconsistencies that a successor was planned for it.

And that successor is OPC UA (Unified Architecture).

FACTORY VS IT AUTOMATION

If you've paid any attention at all to factory automation over the last few years, you've noticed the ever-increasing emphasis on connecting the factory floor to the enterprise. There are many good reasons for this. Some of them are internal: efficiency, productivity, higher quality, and the like. Others are driven by external requirements. Large customers like the Walmart's of the world are demanding higher levels of integration with their suppliers. Regulators are increasing their demands for manufacturers to report on their production processes. Corporate attorneys are "suggesting" that manufacturers archive more data about their production processes.

It wasn't always like this. In the old days (ten years ago?), the production department was a completely separate entity from the rest of the corporation. There was little to no electronic data transfer between the production machines and the company's business systems. Production was a black box. Labor and raw materials went in one end, and finished product came out the other end. Most of the communication was carried out using paper: paper production reports, paper inventory levels, paper raw material usage, paper quality reports, etc. People keyed this information into business systems that used sales data to order raw materials and adjust production levels for the next production cycle – again using paper systems.

Today, the aim is for instantaneous closed-loop communication. As units of product are consumed in the field, that information gets reported back to the machine that made it. The production machine checks its raw material inventory levels and on-hand finished product and schedules more production. It automatically transmits orders for any raw materials it needs from supplier machines. All automatic. All without human intervention.

That's the plan anyway. In practice, it's pretty hard to get there. We don't have the luxury of ripping out all the production machines and replacing them with new, fully integrated machines with high-speed communication mechanisms. Instead, we have to do piecemeal implementations: upgrading and replacing systems one by one as time and funds allow. It's a marathon, not a sprint, to the goal of fully automated systems.

There are many factors impeding progress on our path to fully integrated production systems. Security, of course, is key: the more integrated and connected your production process is, the more vulnerable you are to mischief and worse. Another is the difficulty of replacing fully capitalized and functional systems that are well understood and perfectly operational but lack the integration required for tomorrow's manufacturing vision. And another is the mutual lack of understanding of IT integration by today's control engineers and manufacturing by today's IT people.

The distinction that many people of both stripes miss is that there's a key distinction between the systems on the factory floor and the enterprise. This is the difference between what is called "loosely-coupled" systems and "tightly-coupled" systems. These aren't new concepts, but I don't think they've been examined in the light of the current trend towards the integration of factory floor and enterprise systems.

Factory floor systems can be labeled tightly-coupled. Systems that use Profibus, ProfiNet IO, DeviceNet, EtherNet/IP, or any Modbus version have a very strict architecture. These are really just I/O producers and consumers, despite what some folks at the trade associations might want you to believe.

Let's look at the main characteristics of these tightly-coupled systems:

A Strictly Defined Communication Model – The communication between these systems is inflexible, tightly regulated, and as deterministic as the communication platforms allow.

A Strictly Defined Data Model – The data (really I/O for most of these systems) model is predefined, limited and inflexible.

Strictly Defined Data Types – The data types transported by these systems are limited, predefined and supported by both sides. There is no ability to send data in an open and universal format.

We could look at any of the factory floor protocols, but let's take EtherNet/IP as an example. EtherNet/IP has a very strictly defined communication model. A Scanner uses a very precise communications model in communicating with its Adapters. The Adapters are preconfigured: all data exchanged is predefined, and nothing changes without human intervention. The data exchanged is part of the Adapter's predefined Object Model, and the data is formatted in a way supported by both the Scanner and the Adapter.

Tightly-coupled systems provide much needed, well-defined functionality in a highly specific domain. Expanding operation to other domains or trying to provide more general operation is difficult. Making more generic data and functionality available requires significant programming resources that results in a very inflexible interface.

And that's why tightly-coupled systems are wrong for enterprise communications. That is why I continue to be amused by the proponents of EtherNet/IP and ProfiNet IO as ways to exchange data with enterprise systems. Can they be made to work for a specific application? Yes. But to get there requires a whole lot of effort and results in a difficult-to-maintain, inflexible system that is extremely fragile.

Loosely-coupled systems, on the other hand, provide exactly the right kind of interface for enterprise communications. Loosely-coupled systems decouple the platform from the data, the data from the data model, and provide a much more dynamic mechanism for moving data.

Loosely-coupled systems have these kinds of characteristics:

A Widely Used, Standards-Based Transport Layer – Messages are transported in loosely-coupled systems with open, widely-implemented, highly flexible transports layers: TCP and HTTP.

An Open, Platform Independent Data Encoding – Data is encoded using an open standard data encoding like eXtensible Markup Language (XML) that can be processed by any computer platform.

A Highly Extensible Operating Interface – The interface between loosely-coupled systems is flexible and extensible. SOAP (Simple Object Access Protocol) is the main interface, and it provides a highly flexible mechanism for messaging between loosely-coupled systems.

Essentially, what I've described here is Web Services. Web Services is the backbone of everything we do on the Internet. It is extensible, flexible, platform independent – all required for the ever-expanding Internet.

The challenge is to how to best migrate the tightly-coupled factory floor architectures with the loosely-coupled Web Services architecture of the Internet. It's difficult to migrate today's IO technologies like Modbus TCP, EtherNet/IP and ProfiNet IO. It can be done, but it often results in dreaded, brittle systems that require too much support and cost too much time and money. Integrating these technologies with the loosely-coupled enterprise technologies takes massive amounts of human and computing resources to get anything done. And in the process, we lose lots of important meta-data; we lose resolution and we create fragile and brittle systems that are nightmares to support. And don't even ask about the security holes they create. These systems were not designed to be highly secure. These systems are a house of cards.

And because of the discontinuity between the factory floor and the enterprise, opportunities to mine the factory floor for quality data, interrogate and build databases of maintenance data, feed dashboard reporting systems, gather historical data and feed enterprise analytic systems are lost. Opportunities to improve maintenance procedures, reduce downtime, compare performance at various plants, lines and cells across the enterprise are all lost.

The solution? It's OPC UA. OPC UA can live in both the world of the factory floor and the enterprise.

OPC UA is about reliably, securely and most of all, easily, modeling "Objects" and making those Objects available around the plant floor, to enterprise applications and throughout the corporation. The idea behind it is infinitely broader than anything most of us have ever thought about before.

And it all starts with an Object. An Object that could be as simple as a single piece of data or as sophisticated as a process, a system, or an entire plant.

It might be a combination of data values, meta-data and relationships. Take a Dual Loop Controller: the Dual Loop Controller Object would relate variables for the setpoints and actual values for each loop. Those variables would reference other variables that contain meta-data like the temperature units, high and low setpoints, and text descriptions. The Object might also make available subscriptions to get notifications on changes to the data values or the meta-data for that data value. A Client accessing that one Object can get as little data as it wants (single data value), or an extremely rich set of information that describes that controller and its operation in great detail.

OPC UA is, like its factory-floor cousins, composed of a Client and a Server. The Client device requests information. The Server device provides it. But what the UA Server does is much more sophisticated than what an EtherNet/IP, Modbus TCP or ProfiNet IO Server does.

An OPC UA Server models data, information, processes and systems as Objects and presents those Objects to Clients in ways that are useful to vastly different types of Client applications. And better yet, the UA Server provides sophisticated services that the Client can use, like the Discovery Service.

UA is the future and the perfect technology to bridge the chasm between loosely and tightly-coupled systems.

10 THINGS TO KNOW ABOUT OPC UA

I'm a pretty simple guy. I like steak with a side of mushrooms, not oysters, crepes or pate. I drive an old Buick, not a Mini Cooper convertible or some other foreign car. I watch regular old American football, not arena football or soccer. (I'd make an exception for Victoria Secret football, but I wouldn't be watching for the football.)

I like things simple and straightforward. I like shortcuts. If I was learning OPC UA and wasn't sure if it was something that interested me or that I could use, I would want a shortcut. I'd want a sort of Readers Digest of the most important concepts. I'd want something that would be a starting point so that if I needed to become a proficient I'd have a starting point.

So what follows is a list of the ten most important OPC UA concepts. If all you want is a quick introduction, this chapter is for you.

1. OPC UA is NOT a protocol

It's a common misconception that OPC UA is just another protocol. That couldn't be farther from the truth.

A computer protocol is a set of rules that govern the transfer of data from one computer to another. The protocol specification governs how the data is to appear on the wire, how the conversation between the two computers starts and ends, and what the message structure is between the two computers. It is very rigorous and unforgiving. Now, even though OPC UA also specifies the rules for communication between computers, its vision is more than just moving some arbitrary data from one computer to another. OPC UA is about complete interoperability.

It is an architecture that systematizes how to model data, model systems, model machines and model entire plants. You can model anything in OPC UA.

It is a technology that allows users to customize how data is organized and how information about that data is reported. Notifications on user-selected events can be made on criteria chosen by the user, including by-exception.

It is a scalable technology that can be deployed on small embedded devices and larger Servers. It functions as well in an IT database application as on a recipe management system, on a factory floor, or a maintenance application in a Building Automation system.

It is a technology that provides layers of security that include authorization, authentication, and auditing. The security level can be chosen at runtime and tailored to the needs of the application or plant environment.

OPC UA is a systems architecture that promotes interoperability between all types and manner of systems in various kinds of applications.

2. OPC UA is the successor to OPC (now referred to as OPC Classic)

OPC UA solves the deficiencies and limitations of OPC Classic. In today's world, we need to move data between all sorts of embedded devices, some with specialized operating systems and software, and enterprise/Internet systems. OPC Classic was never designed for that.

- OPC Classic is dependent upon Microsoft – It's built around DCOM (Distributed COM), a technology that is, if not obsolete, certainly being de-emphasized by Microsoft.
- OPC Classic has no support for sophisticated data models – It lacks the ability to adequately represent the kinds of relationships, information, objects and interactions among systems that are important in today's connected world.
- OPC Classic is vulnerable – Microsoft platforms that support COM and DCOM are vulnerable to sophisticated attacks from all sorts of viruses and malware.

UA is the first communication technology built specifically to live in that "no man's land" where data must traverse firewalls, specialized platforms and security barriers to arrive at a place where that data can be turned into information.

3. OPC UA supports the Client–Server architecture

We all very familiar with technologies that have some superior/subordinate relationship. Modbus is one that everyone knows. There is a Modbus RTU Master and a Modbus RTU Slave. A Modbus TCP Client and a Modbus TCP Server. It's the same for EtherNet/IP. Only with EtherNet/IP, the terms are EtherNet/IP Scanner and EtherNet/IP Adapter. The same for BACnet; there is a BACnet Client and a BACnet Server.

The difference between these technologies and OPC UA is that in all of the familiar Industrial and Building Automation protocols, the Client or Master somehow takes ownership of the Server or Slave device. In most of these technologies, once a Client takes ownership of a Slave, no other Client or Master can access it.

That's not true of OPC UA Clients and Servers. In OPC UA, a Server can be configured to accept connections with one, two or any number of Clients. A Client device can connect and access the data in any number of Servers. It's much more a peer relationship in OPC UA, though, like other technologies, Servers simply respond to requests from Clients and never initiate communications.

Another unusual and interesting aspect of this relationship is that in OPC UA, a Server device can allow a Client to dynamically discover what level of interoperability it supports, what services it offers, what transports are available, what security levels are supported, and even the type definitions for data types and object types. A Server in OPC UA is a much more sophisticated device than in many of the technologies you've worked with in the past.

4. OPC UA is a platform-independent and extremely scalable technology

Unlike OPC Classic, OPC UA is designed from the ground up to be platform independent. Ethernet and some mechanism to know the current date/time are the only requirements for OPC UA. OPC UA is being deployed to everything from small chips with less than 64K of code space to large workstations with gigabytes of RAM.

All the components of OPC UA are designed to be scalable, including security, transports, the Information Model and its communication model. Several security models are available that support the level of security appropriate for the device's resources and processor bandwidth.

An encoding mechanism can be selected to provide ease of communication with IT systems (heavy RAM footprint and processor intensive), or one that provides fast, smaller message packets (light RAM footprint and less processor bandwidth).

The same scalability exists with the address space. An address space can be comprised of a few objects with a single variable, or a sophisticated, complex set of interrelated objects.

Transports are often limited in most technologies, but not OPC UA. In OPC UA, a device may only support a single transport, or it may support a set of transports that allow for communications over various kinds of networks.

5. OPC UA integrates well with IT systems

In the last chapter it was noted that there is an ever-increasing emphasis on connecting the factory floor to the enterprise. Management, customers, and suppliers all want increased efficiency, productivity, higher quality, and the like. Government regulation is forcing much more data archiving. All of these factors are increasing the need to connect factory floor systems to non-factory floor processes, enterprise applications, and cloud services.

OPC UA is designed to be well-suited for this. Servers can support the transports used in many traditional IT type applications. Servers can connect with these IT applications using SOAP (Simple Object Access Protocol) or HTTP (Hypertext Transfer Protocol). HTTP is, in fact, the foundation of the data communication used by the World Wide Web.

OPC UA Servers can also support XML encoding, the encoding scheme used by many IT-type applications. It's likely that most Servers in the factory floor automation space will not support XML encoding due to the large resources required to decode and encode XML. Instead, many Servers in that space will support OPC UA Secure Conversation, a more efficient, binary encoding that uses more limited resources.

A key differentiator for OPC UA is that the mapping to the transport layer is totally independent of the OPC UA services, messaging, information and object models. That way, if additional transports are defined in the future, the same OPC UA Information Model, object model and messaging services can be applied to that new transport. OPC UA truly is future-proof.

6. A sophisticated address space model

The address space model for OPC UA is more sophisticated than EtherNet/IP, ProfiNet IO, Modbus or any of the Industrial or Building Automation protocols. The fundamental component of an OPC UA address space is an element called a Node. A Node is described by its attributes (a set of characteristics) and interconnected to other nodes by its references or relationships with other nodes.

All nodes share common attributes with all other nodes in its Node Class. Every Node is an instantiation of its Node Class. Node Classes include the Variable Node Class for defining variables, the Reference Node Class for defining references to other nodes, and the Object Type Node Class, which provides type information for Objects and sets of Objects.

The Attributes of a Node include its Node Class, its Browse Name, its Display Name, its value and its *Node ID*. There are twenty-two possible attributes for a Node. A subset of those nodes are mandatory for each Node Class. For example, only Nodes of the Variable Node Class may instantiate the Value attribute which contains the current data value.

7. OPC UA provides a true information model

The ability of an Object Node to have references to other Object Nodes which further references to other Object Nodes to an unlimited degree provides the capability to form hierarchical relationships which represent systems, processes and information – an Information Model.

An Information Model is nothing more than a logical representation of a physical process. An Information Model can represent something as tiny as a screw, a component of a process like a pump, or something as complex and large as an entire filling machine. The Information Model is simply a well-defined structure of information devoid of any details on how to access process variables, meta-data, or anything else contained within it.

OPC UA provides a systemized way to create and document Information Models. An XML Schema serves as the mechanism for documenting and communicating Information Models. The OPC UA Foundation, trade associations and vendors are all able to develop and propose Information Models for specific devices and systems by describing them using the XML Schema defined by the OPC UA Foundation.

Though OPC UA is not the first organization to systemize the creation and documentation of Information Models, it is the first technology to provide the mechanism to load, transport and reference those Information Models in a running system. A Client device that detects the use of an Information Model in a Server device can access that model (through its URI) and use that model to access the Information Model in a Server.

Many trade groups, including Gas and Oil, the BACnet Association, the PLCopen organization, and others, are using the Information Model processing of UA to define Information Models for their application domains and simply use UA for the standard transports, security and access to their data models.

8. OPC UA is NOT a factory floor protocol

I always cringe when I hear OPC UA compared to EtherNet/IP, ProfiNet IO, or Modbus TCP. That wasn't the case when PROFINET IO came out. I could tell people that it was the equivalent of EtherNet/IP for Siemens controllers. Same kind of technology. Basically the same kind of functionality. Easy to explain.

I can't do that for OPC UA. I could say that it's Web Services for automation systems. Or that it's SOA for automation systems, an even more arcane term. SOA is "Service Oriented Architecture," basically the same thing as Web Services. That's fine if you're an IT guy (or gal) and you understand those terms. You have some context.

But if you're a plant floor guy, it's likely that even though you use Web Services (it's the plumbing for the Internet), you don't know what that term means.

And it's just as likely that if your plant floor guy, you also say, "Why do we need another protocol? Modbus TCP, EtherNet/IP and ProfiNet IO work just fine." The answer is that it's not like EtherNet/IP, ProfiNet IO or Modbus TCP. It's a completely new paradigm for plant floor communications. It's like trying to explain EtherNet/IP to a PLC programmer in 1982. With nothing to compare it to, it's impossible to understand.

In the automation world, the PLC networking paradigm is second nature. You have a PLC; it is a Master kind of device, and it moves data in and out of Slave devices. It uses really simple, transaction-type messaging or some kind of connected messaging. In either case, there is this buffer of output data in a thing called a Programmable Controller. There is a buffer of input data in a bunch of devices called Servers, Slaves or nodes. The buffer of input data moves to the Programmable Controller. The output data buffers move from the Programmable Controller to the devices. Rinse. Repeat. Forever. Done.

OPC UA really lives outside that paradigm. Well, actually, that's not true. OPC UA lives in parallel with that paradigm. It doesn't replace it. It extends it. Adds on to it. Brings it new functionality and creates new use cases and drives new applications. In the end, it increases productivity, enhances quality, and lowers costs by providing not only more data, but also information, and the right kind of information to the production, maintenance, and IT systems that need that information when they need it.

And that's the stuff that EtherNet/IP ProfiNet IO, Modbus TCP and all the rest just can't do.

9. OPC UA is a certifiable standard

Like many other technologies, there is a process to validate an OPC UA implementation. And like many other technologies, there is some certification document that certifies that this device passes the OPC UA certification test suite. But that is where the similarities end.

Due to the scalable nature of OPC UA, it is not possible to have a single test that all devices must pass. Devices that support the Standard Profile, by definition, support many additional features and services than are supported by the Embedded Profile. The Embedded Profile supports more features than the Nano Profile, and so on.

Each OPC UA Profile is composed of small sets of certifiable features. These features, called Conformance Units, are what distinguish one profile from another. The Nano Profile is the smallest profile supporting only the minimum, core functionality required for tiny, embedded devices with no security. The Micro Profile is more comprehensive. It includes the core Conformance Units present in the Nano, plus, in its case, subscription services. The Embedded Profile includes even more Conformance Units, while the Standard UA Profile includes the most Conformance Units.

Certifying scalable systems like this is not a problem in OPC UA. Every OPC UA Server device reports its identification information, including its supported transports, security profile, services supported, and its supported profile. The OPC UA Conformance Test Tool can interrogate a Server to learn its capabilities and specifically test the individual Conformance Units supported in that device.

A successful compliance test results in an electronic Test Certificate being transmitted to the Server. It then carries with it documented evidence of its status as a certified OPC UA device.

10. OPC UA 2015 is still a developing technology

There is a technology adoption lifecycle, and OPC UA is following that cycle. The first systems were really available just a few years ago. Like any other technology adoption, there are the innovators, closely followed by the early adopters. If it's successful, the early majority adopters join in and then the technology reaches its peak adoption.

OPC UA at the time of this writing is still in the technology adoption stage, being deployed by innovators and early adopters. There are some hiccups and false starts that plague any complex development. In the case of OPC UA, it became apparent in 2015 that the Publish/Subscribe services weren't very efficient, and an effort is now being made to enhance that functionality. This is normal and expected as OPC UA continues to become a more and more important factory floor technology.

SHAMELESS PLUG

YOU CAN GET GATEWAYS TO MOVE YOUR FACTORY FLOOR DATA OVER OPC UA FROM REAL TIME AUTOMATION

www.rtaautomation.com/products/

THE UA SERVER

If you've ever worked with Modbus TCP, you know that a Modbus Server device is going to be nearly identical to every other Modbus TCP Server. Because it's a Modbus TCP Server, you could be pretty certain that it exposes most of its data as Modbus registers. And since there are just three function codes that operate on Modbus registers you could be pretty certain you'd know what function codes it supports. Other than the number of data items (registers) in its internal data tables, everything else in it is going to be pretty similar to every other Modbus TCP Server. Similar security (or lack of it), operations, error checking, physical layer and more. You won't find a lot of variability.

All Servers, and we're talking industrial Server devices here, provide the physical interface to the real world. Servers measure physical properties, indicate status, initiate physical actions, and do all sorts of physical measurement and activations in the real world under the direction of a remote Client device. Servers are where the physical world meets the digital world.

OPC UA Servers are fundamentally just like all other Servers in that they are also endpoint devices. OPC UA Servers measure and digitize inputs and transform outputs to their analog equivalents. But that is where the similarity ends. OPC UA Servers are much more sophisticated than Servers of other technologies.

Figure 2 - OPC UA Server overview

You can think about OPC UA Server device like you might think of automobiles; they all have a specific function, but there is a wide range of capabilities and services among them. Dump trucks, buses, and your Buick are all automobiles, but the range of capabilities and services is vastly different.

Profiles

The specific capabilities of an OPC UA Server are described by the Profile it supports. A profile indicates to other devices (electronically) and to people (human readable form) what specific features of the OPC UA specification are supported. Engineers can determine from the Profile if this device is suitable for an application. A Client device can interrogate the Server and determine if it is compatible with the Client and its application and if it should initiate the connection process with the device.

Server Profiles support a wide range of functionality and provide the scalability found in OPC UA. The least functional Profile is the Nano-embedded Server Profile. The Nano Device Server Profile contains the minimum set of functionality needed by small, resource-constrained, embedded devices. The Standard OPC UA Server Profile contains the most comprehensive set of functionality but is usually only suitable for resource-rich devices like PC Servers.

Address Space

All Servers implement some sort of address space. The address space is how a Server device represents its functionality to Client devices. The address space is how the inputs, outputs and data in a device are represented. The Modbus address space consists of a series of 64K blocks. The EtherNet/IP and DeviceNet address space consists of Objects with attributes that represent data points. And the ProfiNet IO and Profibus address space is represented as Racks, Slots, Modules and Channels.

The more sophisticated the address space, the better the device is at representing its data and functionality. Let's imagine a two-channel flow controller with two temperature inputs and two Flow Sensors and look at its representation in a number of different protocols.

Modbus isn't very functional in this flow controller. All you have is a series of registers and no way to provide meta-data regarding the units of the flow values and temperatures and no way to relate the temperature and flow for each channel together.

EtherNet/IP is more functional in that you can have a Flow Channel Object with attributes for the temperature and flow values. Instance #1 of that Object can provide the temperatures and flow for the first channel, while instance #2 would provide the second channel. Meta-data can be provided in other attributes if the flow device vendor chooses to add the additional attributes.

ProfiNet IO is nearly the same as EtherNet/IP except that a module might be used to represent each of the two channels. Additional points could be used to provide meta-data.

In the more functional OPC UA address space, the OPC UA Server represents the data hierarchically in a way that makes more logical sense to a user and lends itself to a visual representation like Figure 3. With OPC UA, you get a much more complete view of hierarchical processes than you do with traditional technologies like Modbus TCP, ProfiNet IO, EtherNet/IP and BACnet.

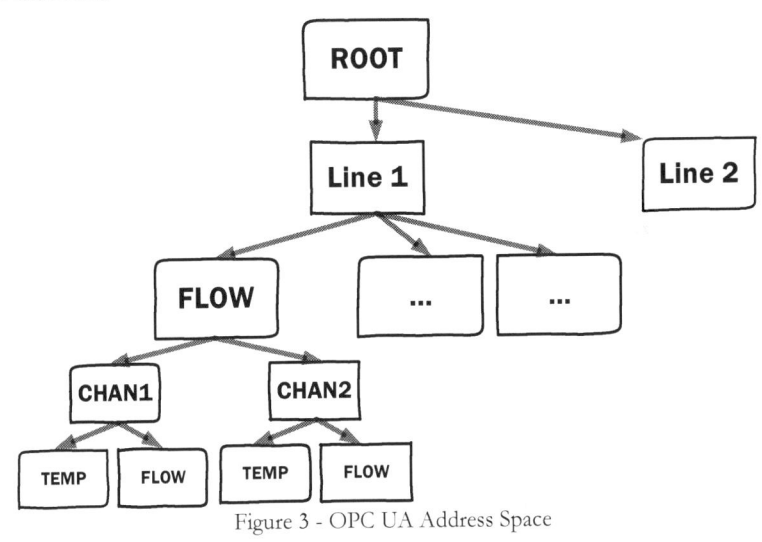

Figure 3 - OPC UA Address Space

A later chapter has a much more complete discussion of the OPC UA address space.

Services

Like other kinds of Servicers, OPC UA Clients issues message requests to a Server and the Server responds with response messages. Message Requests can be directed to the address space of the Server or to the Server's OPC UA communication stack.

Requests issued to the address space can be as simple as reading an Attribute of a Node like the value of the temperature on channel #2 in Figure 3. Or requests can be more sophisticated. A Client can request to browse the address space, add or delete nodes, subscribe to notifications, request historical data, diagnostics, or much more. The specific services supported by a Server are identified by the Profile it supports.

An OPC UA Server implements a much more sophisticated set of services than Servers of other technologies. It announces its availability to interested Client devices, it provides a list of its capabilities and functionality to interested Clients, it provides notifications of different kinds of events, it executes small pieces of logic called methods, it provides address space information in bulk to Clients (Query service), it provides browsing services so that a Client can walk through its address space, and it can allow Clients to modify the node structure of its address space. It can do a lot! And most of that function can't be found in other automation technologies.

Requests can also be directed to the communication stack. These requests can be requests to identify the type and level of security available, what transports the Server supports, to create new connections or sessions, and do other connection management.

OPC UA Server Architecture

The architecture of a Server (Figure 4) can take many forms, but the same components are present no matter how it is architected. Let's take a look at each of the individual components.

The most basic part is the OPC UA stack itself. This is where the heavy lifting of moving messages happens. Incoming messages from Client devices are received from the TCP/IP layer and passed to the OPC UA communications stack. This is where the Client is authenticated, access to the Servers resources are authorized, and where the incoming message is decoded to identify the service requested by the Client. For outgoing messages, this is where the responses are encoded and transmitted to the Client.

THE EVERYMAN'S GUIDE TO OPC UA

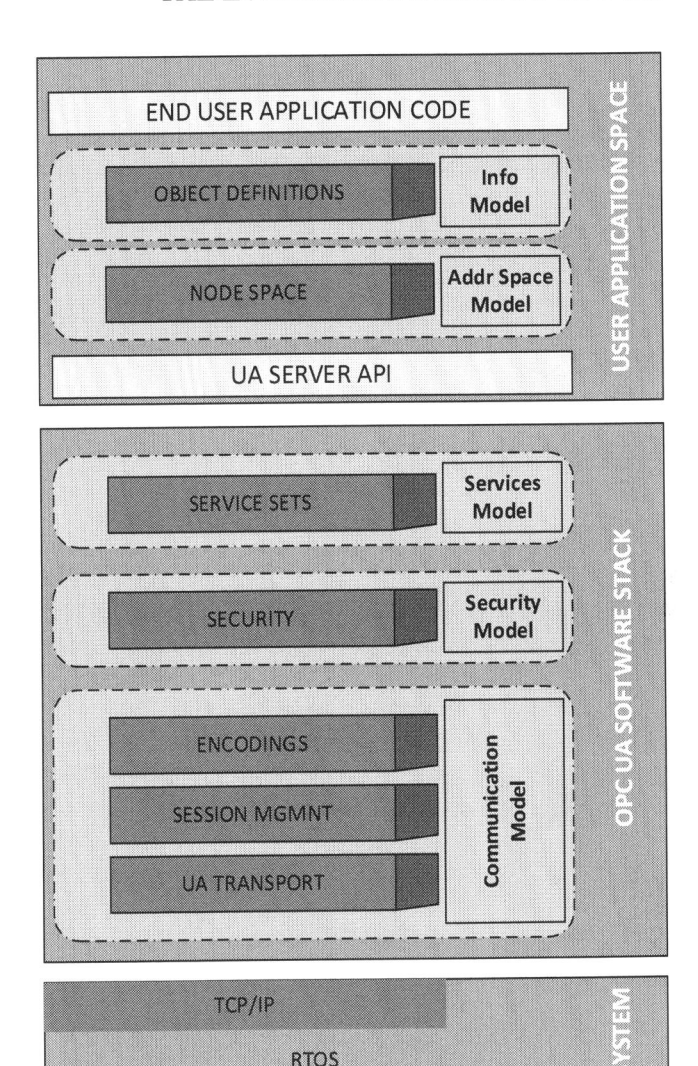

Figure 4 - OPC UA Server software architecture

A service request like "read attribute" is processed in the section of the Server called "service sets." Processing of the request generates a response message, which is passed through the communications part of the stack as described above. That happens for all service requests. The Client request is processed, a response created and then it is encoded, securitized and transported to the Client.

The user application is the part of the architecture that implements the Information Model and the OPC UA address space. This is where the application synchronizes the real world data to the OPC UA address space. Sensor or other data collected from the real world is stored as objects in the

OPC UA address space. Output data in the address space is written to the real world outputs. And in more sophisticated applications, this data is organized to conform to a specific Information Model provided by a trade association or user group of some sort.

Lastly, the final part of an OPC UA Server is the Server end user application. If the device is a motor controller or some other sort of device, this is where the real work of the application happens.

The architecture detailed here is not the architecture that you may find in some of the widely available OPC UA toolkits in the marketplace. Some of these products differentiate the stack processing from any sort of service processing. The service sets, Information Model, address space and the rest are not part of the toolkit. If you plan to purchase a toolkit, make sure that you understand exactly what is included and what is not included.

The Server Object

Most networking technologies provide some sort of mechanism for a networked device to discover the identification information of a device. EtherNet/IP has an Identity Object and other technologies provide similar functionality.

In OPC UA, the Server Object provides that functionality. The Server Object, the black box on the following diagram, fits into the address space model as an Object in the Objects Folder under the Root Object. The Root Object is the top level Object in an OPC UA address space.

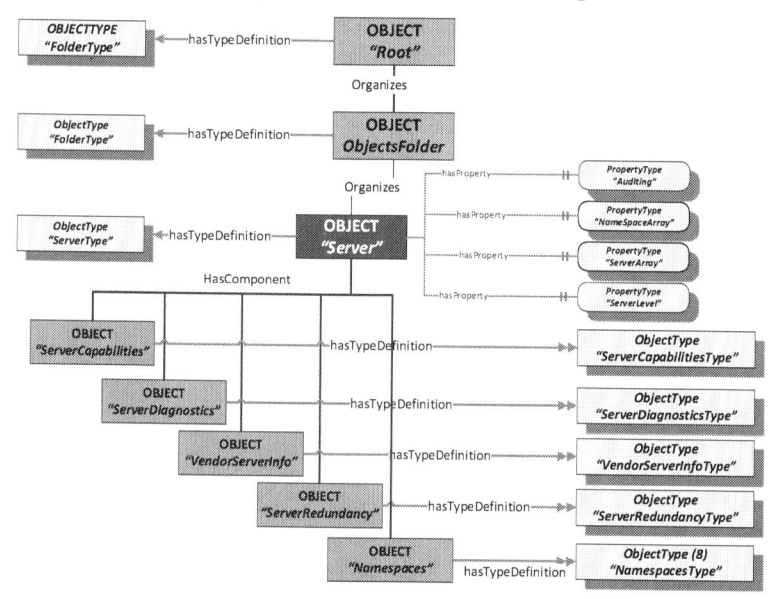

Figure 5 - Server Object in an OPC UA device

In Figure 5, you can see that the Server Object is typed by the *ObjectType ServerType* (all type definitions are lightly shaded boxes). The *ObjectType ServerType* defines a number of variables to include in the Object. These include:

- Server Array – The Server Array is essentially a set of pointers to remote OPC UA Servers that are referenced in the address space.
- Namespace Array – The Namespace Array is a table of URIs for the various namespaces used by nodes in the address space. Index 0 in the array is the OPC Foundation namespace. Index 1 is the local OPC UA Server, while index 2 and above reference other organizations that are responsible for the definition of nodes used in the address space.
- Server Status – Identity and Operational status of the Server, including build information like manufacturer, product codes and software version.
- Service level – A quality of service level from 0 to 255 (best) that Clients can use to judge the relative reliability of Servers in a redundant Server network.
- Auditing – A Boolean indicating if the Server is generating auditing events.

The *ObjectType ServerType* requires the Server Object to support a number of component objects including:

ServerCapabilities

This Object describes the capabilities supported by the OPC UA Server. The *SeverCapabilities* Object contains the list of profiles supported by the Server, the list of signed software certificates from certification testing, the local IDs used for supporting multiple languages, and any number of other operational kinds of data variables.

ServerDiagnostics

This Object contains items like session count, view count, session timeout, subscription counts and lots more counters that assist in troubleshooting Server operation.

VendorServerInfo

The *VendorServerInfo* Object exists to allow vendors to extend Server information by adding additional proprietary information to the Server Object. Vendors can subtype the Object Type and create their own type—"*VendorServerInfo.MettlerScaleInfo*," for example—where they can store additional useful information that is particular to their Server implementation and application.

ServerRedundancy

The *ServerRedundancy* Object describes the redundancy capabilities provided by the Server. This Object is required even if the Server does not provide any redundancy support.

THE NINE MOST IMPORTANT THINGS YOU NEED TO KNOW ABOUT OPC UA SERVERS

Servers are complicated devices with capabilities and operating characteristics that are unlike traditional Servers in Building and Industrial Automation applications. Here's a summary of what you really need to know about OPC UA Servers:

1. The specific capabilities of an OPC UA Server are described by the Profile it supports. The Nano Profile is the least functional, smallest profile, while the Standard Profile contains its full, complete capabilities. Every part of the Server's architecture is a function of the Server Profile selected by the Server device vendor. The Profile identifies the transport: Web Services, HTTPS, OPC UA Binary TCP, or something else. The Profile identifies what the type and level of security that is used and the specific services that are supported.

2. All Servers support a Discovery Service that Clients use to discover OPC UA Servers that are available in a system. It communicates the transports, security and other communications characteristics supported by that specific Server. A network may also contain a special Server, called a Discovery Server that aggregates the information supplied by all the Servers on a network.

3. Once a connection is made, a Client that is not familiar with the Server's operating characteristics reads the Server Profile to understand what certified functionality is incorporated in the Server, what encodings it uses, its support for subscriptions and notifications, what other services are supported, and much more.

4. Servers are only required to support a single transport. Any one of several transports can be supported by a Server. Clients, on the other hand, must support multiple transports.

5. A Server can support either XML or binary encodings, or both. The vendor building the Server makes that decision at design time.

6. The Server Object, which is not required for the low end Nano Profile, provides the vendor specific information about the Server.

7. A Server represents data to Client devices through its address space. An address space is almost always built at design time, but there is no reason why it can't be modified during Server operation. An address space consists of nodes defined by attributes that reference other

nodes.

8. Servers provide the type definitions for Objects, Variables, and Data Types that are not predefined by the OPC UA Foundation. Some type definitions may not be typed locally if they are defined in another Server or by some other organization. In these cases, instead of the actual type definition, there is a reference to where the type definition can be found.

9. The Publish/Subscribe functionality of OPC UA is supported in several different ways. Check with your Client and Server vendor to ensure compatibility.

OPC UA Server devices are end point devices that are more capable than Servers of other industrial technologies. UA Servers are designed to function in a wide variety of applications from low end sensor/actuator devices to sophisticated instruments and factory floor automation devices. The functionality of a Server is described by the Profile it supports. The least capable Servers use the Nano Profile while the most capable Servers use the Standard Profile. No matter how capable the Server, it connects to OPC UA Client devices, the subject of the next chapter.

THE UA CLIENT

In most industrial networking technologies, there is a controlling device: a device that connects to and controls one or more end devices. In ProfiNet IO, this device is known as an IO-Controller. In Modbus TCP, it is a Client. In EtherNet/IP, it's called a Scanner, and a Master in DeviceNet and Modbus RTU. No matter what it is called, that device is configured by the user to open connections with one or more end devices, send outputs to that end device, receive inputs from that end device and, occasionally, send asynchronous command requests.

In OPC UA, a device of this type is known as an OPC UA Client. Like controlling devices in these other technologies, an OPC UA Client device sends message packets to Server devices and receives responses from its Server devices. But beyond this basic functionality, an OPC UA Client device is fundamentally more sophisticated than controllers in other technologies. The extended functionality included in an OPC UA Client includes:

- The use of the OPC UA Discovery process to find eligible Servers on local and remote networks.
- The ability to identify a Server's application type, application name, and set of hostnames[1] whenever a new Server is discovered.
- Messaging Servers on special non-operational endpoints[2] that exists to provide information such as the security mode (signed or encrypted messages), the Server's security policy, the transports it supports, and the Servers Application Instance Certificate.

---

[1] Hostnames in this context are simply human-readable nicknames that represent an IP Address, Port Number and name.

[2] Endpoints are how web devices make resources available to Client devices. On the web, a service endpoint is an entity, processor or some other resource that Clients can address to gain access to that resource. In OPC UA there are generally two endpoints: one unsecure endpoint that allows Clients to identify Servers, and another secure endpoint used for operational access.

- Creation of authenticated connections with the Server using the security mode supported by the Server (or an unsecure connection if the Server doesn't support security.)
- Creation of secure and authorized logical communication sessions with Servers using the security mode supported by the Server.
- Evaluation of the capability of the Server to perform required services for the Client application by inspecting the Server's Software Certificate[3].
- Sending message requests to Servers requesting the execution of one or more of the service requests supported by the Server.
- Configuring the Server to notify the Client of alarm conditions, program outputs and data changes.

One of the unique aspects of the Client/Server relationship in OPC UA is that there is no restriction on how many Clients can talk to one Server, how many Servers can be connected to a single Client, or even how two Servers can communicate. No matter what the configuration, Clients send message requests to Servers, and Servers respond with responses and notification requests as shown in Figure 6.

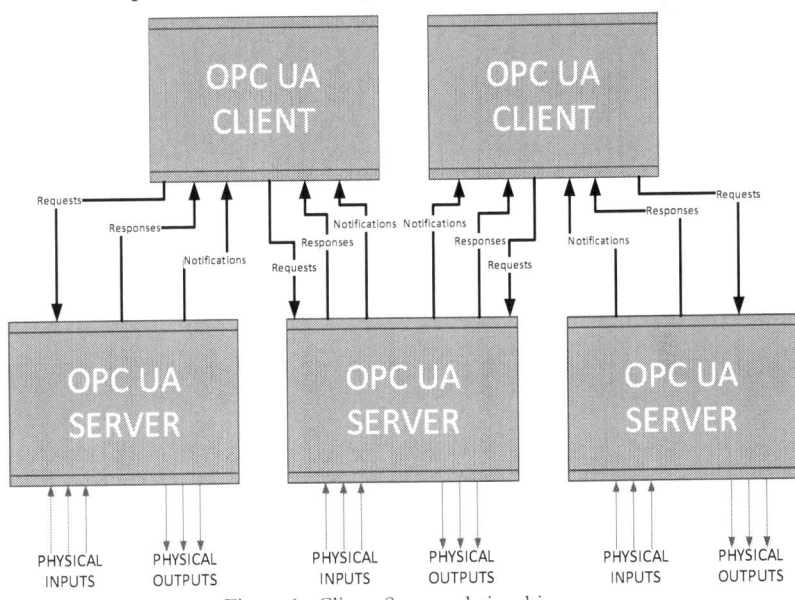

Figure 6 - Client–Server relationships

In this diagram, two Clients are each connected to two Servers with the middle Server connected to both Client devices. Any sort of connection diagram is possible, as some Servers embed a Client and can function as either a Client or a Server or both.

---

[3] A Software Certificate lists the specific functionality validated during the OPC UA Certification testing of the Server. OPC UA groups functionality into small, testable items called Conformance Units.

How OPC UA Clients Find OPC UA Servers

In most industrial and building technologies, you can't search for Servers on the network. Normally, some "out-of-band" configuration occurs using a dedicated vendor specific tool or a configuration web page. A user configures the controlling device with the address of the Server and the identifying data it needs to connect to and transfer data with the Server.

Unlike these technologies, OPC UA contains a built-in Discovery process where Clients can find and connect to Servers without user intervention. The Discovery process is field configurable by the user, as some installations may not want to expose their Servers, and other installations may want to restrict a Server to communications with specific Client devices.

Some installations may use a Discovery Server to catalog available OPC UA Servers. Clients can get information on available Servers by accessing the Discovery Server instead of the individual Servers. When the Discovery Server is integrated on the same platform as the Server itself, it is known as a Local Discovery Server (LDS). When the Discovery Server is able to catalog Servers supported over multicast networks, it is known as a Local Discovery Server with Multicast Extension (LDS-ME). When the Discovery Server catalogs available Servers in its address space, it is known as a Global Discovery Server (GDS).

Every Client has its own process for discovering Servers, but there is a set of alternatives that an OPC UA Client can use:

1. Get the Server's information directly from the Server. Get the hostname of the Server from the user at the installation using some out-of-band mechanism.
2. Get Server information from an LDS on the same host as the Client.
3. Get Server information from external LDS Servers. The hostnames for these Servers would be obtained from the user at the installation using some out-of-band mechanism.
4. Get Server information from external LDS-ME Servers. The hostnames for these Servers would be obtained from the user at the installation using some out-of-band mechanism.
5. Search the address space of a GDS Server to get Server information.

The primary service for discovering Servers is the *Find Servers* service. A *Find Servers* request directed to a non-Discovery Server returns the application description of that Server. A *Find Servers* request directed to a Discovery Server returns the application description for all the Servers that match the criteria specified in the request.

The *Find Servers On Network* request performs the identical function on a LDS-ME Server and returns the application descriptions for all the

Servers that match the criteria specified in the request.

From the Application Description the Client retrieves:

- The Device type (Discovery Server, Client, Server, or Client & Server)
- The ASCII name for the device such as "Delco Inc. 32bit Valve Block
- The Discovery Endpoint for the device, which is part of the application URI. For an embedded device it might be something similar to:
    "192.168.0.100:realtimeautomation:modbus gateway"
- A unique product URI which uniquely identifies the product among all other products from that vendor

How Clients Access Server Devices

Once a host has identified a Server of interest and has the Discovery endpoint and product information, its next step is to identify what transports and security that device supports. To get that information, the Client issues the *Get Endpoints* request. The response to that request is an array of structures called Endpoints Description.

Each structure describes an endpoint in the Server. The Endpoint Description consists of:

- The Application Description for the Server – the same Application Description described earlier
- The Server's Application Instance Certificate – the X509 Certificate for the Server. The Application Instance Certificate is the certificate received from a CA (Certification Authority) that authenticates the Server application for the Client
- The Server's Security Mode – one of "all messages must be signed," "all messages must be signed and encrypted," or "messages are not signed or encrypted" on this endpoint
- The transport supported on that endpoint: one of HTTPS, HTTP, UATCP, or some other supported transport

The Client device examines all the endpoints available in the Server and selects the one best suited for its application. A web Client operating in the enterprise environment might choose an endpoint with HTTPS and signed and encrypted messages. A Client inside the firewall of a manufacturing facility might choose performance over security and choose and endpoint with the UATCP transport, binary encoding, and no security.

How A Client Connects To A Server

Once an endpoint is chosen, the OPC UA Client must make three connections before it can begin to issue service requests to a Server. First, the Client must make a connection to the Server using the transport assigned to that endpoint. Today the transport options are HTTPS, HTTP, and UA TCP, but that list is going to expand as time goes on.

Second, the Client must establish a secure communication path between itself and the Server device. This connection, a "channel" in OPC UA terms, is a long running, secure, authenticated connection between the Client and the Server. The channel connection is a device-to-device communication path. The channel authenticates each side in that communication path and exchanges the keys needed for secure communications between the Client and Server. The keys to encode and decode the secure messages are specific to the security profile implemented on the selected endpoint.

Certificates signed by a Certification Authority (CA) are used by the channel to validate the identity of each device. The Client sends its Client Certificate, signed by the Certification Authority (CA), to the Server, and the Server sends its Server Certificate, also signed by the Certification Authority (CA) to the Client. The signed certificates provide proof that both Client and Server are the devices they claim to be.

Even though the Server may authenticate the Client, it can still reject the Secure Channel request if the Client is not approved for connection to the Server. Note that this approval is an application approval and outside the scope of OPC UA. For example, some Servers may allow an administrator to enter a list of approved Clients that can connect to the Server. That's Server-specific functionality.

Once the channel is authorized, the final step is for the Client to establish a logical connection with the Server called a session. The session is a long term, authorized connection between two applications. Sessions are not active until enabled by a Client request. Once enabled, a session can remain open even if the underlying Secure Channel is closed. In that case, another channel can be created to host that open session. In fact, the Client can access the session from any channel. Sessions typically have a lifetime, and the Client must renew a session before the "homeless" session's lifetime expires and it is deleted.

Where the channel processing serves to authenticate the devices operating over the connection, sessions serve to authorize the Client application to access the Server application. Just as with the channel, the Server can reject the Create Session request. Servers may have numerous reasons for rejecting a Client's application's access to the Servers address space, but that approval or rejection is application dependent and outside the scope of OPC UA.

UA Clients – What You Need To Know

There are eight concepts that are important to remember when thinking about OPC UA Clients:

1. Client devices request services from OPC UA Server devices. Server

devices send response messages and notifications to the OPC UA Client device.
2. The Subscription Service Set, which drives notifications, and the Read Service of the Attribute Service Set are the primary services that OPC UA Clients use to interact with the address space on an OPC UA Server.
3. Clients find OPC UA Server devices in multiple ways. Clients can find Servers using traditional configuration, by using a Local Discovery Server, by using a Local Discovery Server with a Multicast Extension, or by using a Global Discovery Server.
4. Once a Client finds a Server, it obtains the list of available endpoints and selects an endpoint that supports the security profile and transport that matches its application requirements.
5. Clients begin the process of accessing an OPC UA Server by creating a channel, a long term, connection between it and an OPC UA Server. Channels are the authenticated connections between two devices.
6. Once the channel is established, Clients create sessions, long term, logical connections between OPC UA applications. A session is the authorized connection between the Client's application and the Server's address space.
7. Clients can subscribe to data value changes, alarm conditions, and any results from programs executed by Servers. Servers publish notifications back to the Client when those items are triggered.
8. Clients invoke methods, which are small program segments. Programs can return results to the Client in the Method call or in a Notification if the Client subscribes to it.

With this short background on Clients and Servers, we can turn our attention to the fundamentals of OPC UA communications, beginning with OPC UA transports.

INTRODUCTION TO SECTION II

The next chapters describe specific features of the OPC UA architecture. The chapters are organized to provide the reader with as little or as much detail as preferred:

- A "What's Interesting" section of each chapter provides a quick overview of that attribute of the technology.
- A "What You Need To Know" section provides the essential details for readers who wants a quick summary of the items that they should know about the topic.
- For readers who desire a much deeper dive with a lot of detailed information, there is a "What Are The Details" section.
- When there is something important about that aspect of OPC UA technology that provides differentiation for it over other technologies, that is described in a "What's the OPC UA Advantage" section.

Since these chapters provide readers with alternative reading paths, the reader who reads a chapter from beginning to end, will unavoidably find duplication of material from section to section.

TRANSPORT LAYERS

What's Interesting?

What's interesting about OPC UA Transports is the variety and flexibility that's provided. Server devices can support a low end, high throughput, insecure, transport layer, or a highly secure, certificate based, processor intensive transport. It's entirely up to the Server designer which transport to choose and how many different transports to support, as they are not limited to a single transport.

Client devices, on the other hand, need to be much more sophisticated. A Client device must not only support all the different types of transport layers, to ensure that they can access any kind of Server they wish, but they also must have the ability to interrogate a Server (or an aggregating Server) about what kind(s) of transports are supported.

What Do You Need to Know?

Once an OPC UA application forms a UA message or a response, it must send it somewhere. Transports are the low-level mechanisms for moving those serialized messages from one place to another.

Transports in OPC UA are not limited. OPC UA operates in a very broad technology space, and UA devices can support multiple transports or even custom or proprietary transports. OPC UA devices can be anything from a factory floor sensor or actuator to a Programmable Controller, a Human Interface Device, a Windows Server operating a massive Oracle database, or an undersea pipeline controller. A rich set of support transports are required to support the OPC UA mission of being a completely scalable solution.

The UA specification defines a number of transports that Clients must support. A quick overview of each follows.

SOAP / HTTP TRANSPORTS - HTTP (Hypertext Transfer Protocol)

is the connectionless, stateless, request-response protocol that you use every time you access a web page. SOAP (Simple Object Access Protocol) is an XML messaging protocol that provides a mechanism for applications to encode messages to other applications.

HTTPS TRANSPORT - Hypertext Transfer Protocol Secure (HTTPS) is the secure version of HTTP. It means that all communications between your browser and that website are encrypted. Just as with HTTP, SOAP is used as the request-response protocol to move the OPC UA requests between Clients and Servers.

UA TCP TRANSPORT – UA TCP is a simple TCP-based protocol designed for Server devices lacking the resources to implement XML encoding and HTTP/SOAP type transports. UA TCP uses binary encoding and a simple message structure that can be implemented on low end Servers.

What's the OPC UA Advantage?

The advantage to OPC UA is that it is the first information technology that can easily interoperate at both the machine control communications and the enterprise levels. Most other technologies are designed for a specific application. OPC UA is more flexible in that it provides connectivity advantages for all levels of the automation architecture.

This isn't to say that OPC UA is "better" than EtherNet/IP, ProfiNet IO, or Modbus TCP. For moving IO data around a machine, these protocols provide the right combination of transports, functionality and simplicity that enable machine control with a networked I/O. They are very good technologies for the machine control level of the automation hierarchy.

What is disputed, however, is how well these technologies can be adapted to communicating with the enterprise, even though the trade organizations that promote these technologies argue that they are perfectly adaptable. The simple fact is that all networked factory floor communication protocols are optimized for the factory floor and are not very well-suited for moving information between a machine cell and an enterprise application. They lack the data representation, data formats, communication interfaces, and services that adapt well to IT-like kinds of enterprise applications.

What makes OPC UA transports so powerful is that OPC UA requests to Read or Write an Attribute can use standard Web Services technologies. UA requests and responses can be encoded as XML, placed inside a SOAP request, and transferred to an IT application that already knows how to handle the HTTP, the SOAP message, and the XML. OPC UA has an essentially "free" mechanism for sending messages between the factory floor and IT devices.

What Are the Details?

Many of the common Industrial Automation (IA) protocol technologies limit the available transports. There is good reason for this. IA devices operate in a very organized, well-defined technology space with devices that are designed to easily integrate with each other. There is no need for a wide selection of transports.

In Industrial Automation, devices that want to communicate with Programmable Controllers must use the transport that is defined for the communication technology supported by that brand of Programmable controller (PLC). For Rockwell PLCs, that means TCP, UDP and IP, the transports for EtherNet/IP or CAN (which is the transport for DeviceNet). For Siemens PLCs, it means TCP and IP, the transports for ProfiNet IO. Other PLCs use other transports or simply transmit the application layer protocol using RS232 or RS485 electrical signaling.

The transports of OPC UA are not limited like this. The technology space where OPC UA operates is much more extensive and requires support for many different transports and the capability to add new transports in the future. Transports are about how an OPC UA message is moved from your node to some other node on the network. When we discuss transports, we don't talk about the message content, how data is secured, how a node asks another OPC UA node for information, and so on. Instead, we discuss the low-level mechanisms for getting the OPC UA packet from "here" to "there."

The Internet has a set of mechanisms that are pretty universal and make it easy for different computers on the web to move data. I can, for example, build a website to sell dog leashes. Undoubtedly, they would be the best dog leashes in the world and you would buy one for every day of the week. But while you are browsing my dog leash web site, you would find a bunch of books from Amazon for training your precious little Fluffy, which you could also buy from me.

The reason that happens is because, on the Internet, there are well-defined sets of protocols and web interfaces making that sort of seamless communication possible. It's what makes the web so incredibly valuable as we not only buy books on how to make Fluffy roll over, but also stay in touch with Aunt Sophie and rate the crème brulee we ate last night.

The plumbing, what goes on behind the scenes, and makes this all happen is new to us, as Industrial Automation guys. But if we're going to work in a world where IT in effect runs the factory floor, and wants a lot of information from our machines, we have to understand how they work.

One of the first things you'll have to learn is SOA. That stands for

Service Oriented Architecture. It's a fancy term for nodes on the web, where a computer can come and ask that node a bunch of questions. A computer can ask questions like:

Who are you?

What kinds of data do you have?

How can I get your data?

It's pretty nice, and when implemented, it's called Web Services. It means that there is a standard way of how computers on the web talk to one another. It's the "language" of the web.

But the key thing to remember here is that it's not the "phone line," it's not the physical connection. It's the difference between the mechanism that makes the connection and the language that is spoken over that connection. For example, I can call Korea and get a connection even if I can't speak a word of Korean.

So what are the connections adopted by OPC UA that make the web so universally useful? Let's look at them in more detail than we did in the previous section:

SOAP / HTTP Transports

HTTP (Hypertext Transfer Protocol) is the connectionless, stateless, request-response protocol that you use every time you access a web page (the "http" in http://www.rtaautomation.com/). HTTP is the standard way that a Client can request data like HTML files, images, and query results from a Server.

SOAP (Simple Object Access Protocol) is an XML messaging protocol that provides a mechanism for applications to encode messages to other applications. An application can encode a SOAP request that asks another application to perform a service or return data.

When used as an OPC UA transport, OPC UA requests and responses are encoded as SOAP requests. These can be easily decoded by standard mechanisms of many IT-type applications. What makes this so powerful is that OPC UA requests to Read or Write an Attribute can be encoded as XML, placed inside a SOAP request, and transferred to an IT application that already knows how to handle the HTTP, the SOAP message, and the XML. OPC UA has gained an essentially free mechanism for sending messages between factory floor and IT devices.

HTTPS Transport

Hypertext Transfer Protocol Secure (HTTPS) is the secure version of HTTP. When you see a web URL that is preceded by "https:" it means that all communications between your browser and that website are encrypted. Just as with HTTP, SOAP is the request-response protocol for moving the OPC UA requests between Clients and Servers.

There's another secret to easily moving data on dog books from one node to another—that's XML, the encoding layer. This is how data is formatted so it can be understood by each side.

A long time ago, it was nearly impossible for two computers to pass data back and forth. Some computers put formatted data, Most Significant Byte (MSB), first. Others put the Least Significant Byte first (LSB). There was a lot of byte swapping and other shenanigans that needed to happen before two computers could pass a little data from one to another.

That's why XML was invented. XML is an encoding mechanism that sends data in ASCII format. ASCII, encodes data using standard ASCII tables, encoding mechanisms that all computers understand. XML encodes data with some sort of name, the ASCII value, and a terminating name. For example:

<DogLeashX200Color>Brown</DogLeashX200Color>
<DogLeashX200Price>$10.95</DogLeashX200Price>

Transport and encoding mechanisms work great if both the Client and the Server are located on your enterprise network. If you have a Windows system, all of this is built-in. It's done without even thinking about it.

Transports for Industrial Automation

If you have a computer with a ton of memory and a really hefty processor, you can easily store the XML as it is received, convert it, and process it. However, there aren't many devices that support these kinds of transports within your manufacturing network; as most embedded devices—like a glue controller, valve block, motor drive, Programmable Controller, and the like—don't have Windows.

We need a different encoding system and a different transport: one that is friendlier to resource-limited devices like we have on the factory floor. How does OPC UA handle that need? Well, UA designed an encoding and transport for manufacturing devices, one that could be supported by embedded devices with limited memory and processing power.

Instead of XML encoding, OPC UA provides a binary encoding for these devices. This binary encoding vastly reduces the amount of bits that have to be transferred from one device to another. It does this by sending the bits as a well-defined set of ones and zeros (without the overhead you have in XML encoding). It's fast, efficient, and makes low resourced devices functional on the network.

Instead of the HTTPS and Web Services transports, UA provides a different transport for moving data from one place to another. It's called UA TCP, because it's specific to UA and uses standard TCP for the connection layer.

All Clients and Servers use UA TCP. UA TCP is simple TCP-based protocol designed for Server devices that lack the resources to implement XML encoding and HTTP/SOAP type transports. UA TCP uses binary encoding and a simple message structure that can be implemented on low-end Servers.

Hints for Reading the OPC UA Specification

Part 6 of the OPC UA specification is named "Mappings," to describe how OPC UA protocol messages are mapped to the message security and transport layers of OPC UA. It describes the two encodings used in UA (XML and Binary), the various ways that messages can be secured (or not secured), and the three transports currently available in UA; HTTP/SOAP, HTTPS, and OPC UA Secure Conversation.

Most designers of embedded devices who desire security will want to focus on the discussion of OPC UA Secure Conversation (UASC). It's important to note in this section that Opening and Closing of Secure Channels is different than the Open and Close channel requests, described separately in the services section of the specification (Part 4).

What is not explicitly clear in Part 6 is this; the service Request or Response structure carried by transport layers and encoded with XML or binary encoding is also encoded as an Extension Object. Extension Objects are containers for complex sets of data types, containing any of the OPC UA built-in types. The fact that Server Requests and Responses use the Extension Object structure can be found in Part 6, but you have to look to find it.

You will also want to examine Part 6 closely in order to understand the details of the UA TCP transport layer, described only briefly in this chapter. You will want to understand the various low-level messages that are implemented by UA TCP, and how connections are established.!

If you're new to TCP/IP, you can get a TCP/IP Protocol Guide. Just visit either of the following web pages:

http://www.rtaautomation.com/technologies/opcuainfo/

DEVICE DISCOVERY

What's Interesting?

What's interesting about OPC UA Discovery is that there is a very flexible and sophisticated mechanism for Client devices to find and identify Server devices. The traditional mechanism for matching a Client (also called an Initiator or Master device) to a Server (or a Target or Slave device) is for the user to identify the Server to the Client during some sort of configuration process. These mechanisms are effective for the kinds of factory floor applications we've used in the past. In fact, for low-level sensor/actuator networks, they would still be recommended. But OPC UA is designed to function equally well on the factory floor and the enterprise, so what's really interesting is how OPC UA created a more robust and flexible mechanism that works equally well on the factory floor with embedded devices as it does in the enterprise with IT devices.

OPC UA designed Device Discovery using the Service Oriented Architecture (SOA) model. What makes SOA so interesting, and the backbone of a lot of Enterprise and Internet applications, is that there is a standard mechanism for Clients to discover Server devices, interrogate them to see what services they offer, and connect to them. OPC UA extends the SOA model to define a process in which Clients can find Server devices that choose to reveal themselves, interrogate them to determine the various ways that the Client can interact with them, and determine what capabilities they have that might be of interest.

Device Discovery is designed to support the various kinds of architectures that automation people might use. There are adaptations of it for the simplest One-Client-to-One-Server on a local host to applications where multiple Clients and Servers are operating in a local network, to highly distributed applications that operate across the Internet.

What Do You Need to Know?

There is a term that is many of us in Industrial Automation have heard but may not fully understand. That term is endpoint. An endpoint is a connection to a device that offers some specific functionality that is sometimes only available through that specific connection. For example, you can think of the programming port on a Programmable Controller as an endpoint. It uses a specific physical layer with a specific transport to accomplish downloading a program. That programming port is an endpoint, and ifs dedicated to programming, it will differ from every other communication port (endpoint) on the Programmable Controller.

Typical Ethernet devices in Industrial Automation have a single endpoint. ProfiNet IO devices have an endpoint that supports ProfiNet IO connections with cyclic and acyclic message transfer. EtherNet/IP devices have an endpoint that supports CIP (Common Industrial Protocol) messaging. Modbus TCP devices have an endpoint that supports Modbus Messages over Ethernet.

The only real cases where devices we are familiar with in Industrial Automation have multiple endpoints are those devices that support multiple protocols. If a device supports EtherNet/IP and Modbus TCP, then there will actually be two: http://192.168.0.100:502/ for Modbus TCP and http://192.168.0.100:44818/ for EtherNet/IP. That hasn't been important to us in the past because a Modbus Client or an EtherNet/IP Scanner device (Client/Initiator) "knows" to send messages to a Modbus Server or an Adapter device (Slave/Initiator) using those endpoints.

But now, with OPC UA, endpoints are not only much more important, they are much more sophisticated. OPC UA devices can have any number of endpoints. Some may have only one or two. Others might have as many as five or ten. Some endpoints will use HTTP, HTTPS, or UA Secure Channel as the transport. Some endpoints will require signed messages, others will require signed and encrypted messages, and still others may not use any security at all. Some endpoints will exist for particular purposes, such as the Discovery endpoint, which Clients use to get the endpoint information. Those endpoints will usually not provide the functionality that you might find on other endpoints.

To understand OPC UA Device Discovery, it is important to understand the basic steps that are used in the Device Discovery process where an OPC UA Client device finds, interrogates and connects to OPC UA Sever devices:

10. A Server powers on, and if it chooses to provide its own Discovery

Service support, it opens its Discovery port for messages from Client devices that want endpoint information.

11. If a Server chooses to let another Server provide Discovery Services, it registers itself with Local Discovery Servers (LDS) or multicast Discovery Servers (LDS-ME). The Discovery Servers may be resident on the same platform as the Server or on a platform someplace else on the network. Note that some Servers may choose to remain private and only be available to Clients that are configured to know about them.

12. A Client that finds a Server in an LDS can retrieve the application description for the Server and information on accessing the Server's Discovery endpoint where it can get more detailed information on the Server.

13. If a Client is interested in connecting to a Server, it uses the Server's Discovery endpoint to get the list of endpoints supported. The information on each endpoint includes the transport and security it supports.

14. If the Client finds an endpoint that meets its application requirements, security requirements and transport requirements, it begins the connection process with the Server on that endpoint.

What Are the Details?

The traditional mechanism for matching Client (or Master/Initiator) up with Server (or Slave/Target) is for the user to manually identify the Server to the Client. In a closed system, the Server address on the network is well known and never varies. The identification is hard coded in the Client's software. In other, more generic systems, there is a front panel, web page or message set that a user or installer can use to identify the Slave to the Client.

These mechanisms are effective for the kinds of factory floor applications we've used in the past. In fact, for low-level sensor/actuator networks, they could still be recommended. But OPC UA is designed to work equally well on the factory floor and the enterprise, so a more robust and flexible mechanism is required.

To play well with Enterprise and Internet-type applications, OPC UA had to comply with the rules of the game in the IT world. And though there are a lot of standards, you can categorize how many enterprise applications communicate under the title of SOA – Service Oriented Architecture.

SOA is a communication architecture in which applications provide services to each other. It says nothing about transports, encodings, what protocols are offered or anything else. SOA just means that Server devices provide services that they can perform for Clients. Client devices connect (if allowed), use one or more of the services for some period of time, and then disconnect. Clients connect when they need the service and disconnect when they don't. SOA provides a much more loosely-coupled architecture than the always-connected Industrial Automation technologies like DeviceNet, EtherNet/IP, Profibus or ProfiNet IO.

OPC UA is designed on the SOA model. One of the interesting features of the OPC UA implementation of SOA is that there is a sophisticated mechanism for Clients to discover Server devices, interrogate them to see what services they offer, and connect to them. The implementation is designed for all sorts of network architectures from the simplest local network with one Client and one Server to the largest device network with thousands of devices connected over the Internet.

Endpoints

As discussed earlier, a key term we need to know in connection with the OPC UA Discovery process that many of us in Industrial Automation may not really be familiar with is "endpoint." Endpoints are commonly used in IT applications, but it's not a term that we regularly use with our EtherNet/IP, ProfiNet IO, and Modbus TCP applications.

An endpoint is nothing more than an access point into a device with a specific set of capabilities. In most Industrial Automation devices, there is just a single entry point, something like 192.168.0.100. All device services are provided at that same address.

Devices that support more than one Ethernet application layer can be thought of as having two entry points. For example, one entry point for EtherNet/IP, 192.168.0.100:44818, and one entry for Modbus TCP, 192.168.0.100:502, where the number after the colon is the port number assigned by IANA (Internet Assigned Names Authority) to that application layer. Device vendors can create endpoints to support specific services. An endpoint like 192.168.0.100:502/diagnostics might support some special diagnostic service in a Modbus TCP device, or "/logic" might be a port assigned for downloading of logic.

OPC UA Server devices usually support at least two endpoints: one endpoint to support the Discovery Services described in this chapter and another endpoint to support Server operation with a specific transport, encoding and security profile. Some may only support a single endpoint and disable the Discovery endpoint. Any number of endpoints may be supported to support all the combinations of transports, encodings, and security profile supported by the Server.

```
┌─────────────────────────────────────────────────────┐
│           OPC UA SERVER WITH THREE ENDPOINTS        │
│    Discovery – Discovery Endpoint with No Security  │
│  IT Interface – Operational Endpoint with HTTPS Security │
│   Embedded Interface – OPC UA Secure Conversation with │
│            Encryption & Certificate Security        │
└─────────────────────────────────────────────────────┘
```

EMBEDDED ENDPOINT 192.168.0.100:xxxx/embedded

IT COMMS ENDPOINT 204.15.5.0:4840/enterprise

DISCOVERY ENDPOINT 204.15.5.0:4840/UAdiscovery

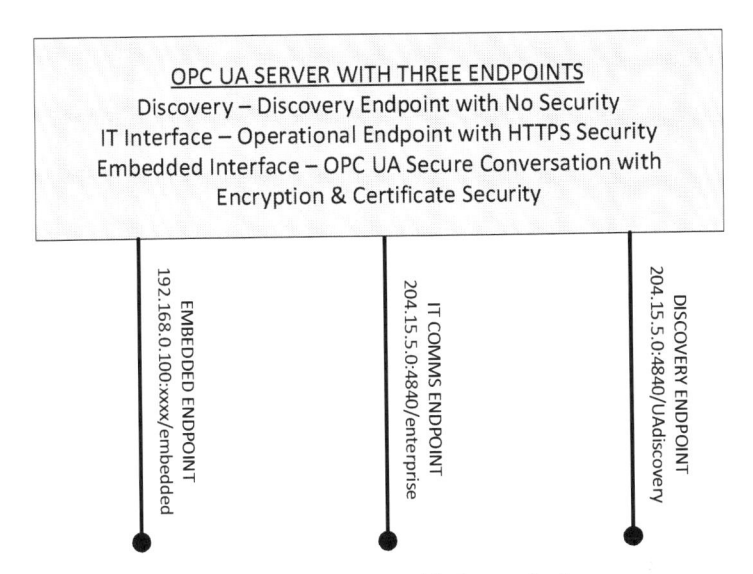

Figure 7 OPC UA Server with three endpoints

Figure 7 is an example of an OPC UA device with three endpoints. In this example, there are two different TCP/IP addresses used. The device exists on both an IT network and an embedded network. The Discovery and IT endpoints use the same NIC (Network Interface card) and have the same address. The embedded interface uses the NIC and TCP/IP address for the manufacturing network. Note that in this example, the embedded endpoint does not use the standard 4840 OPC UA port. OPC UA devices are not required to use Port 4840.

The Discovery endpoint provides the Discovery Services. Like all Discovery endpoints, it is unsecure. The security policies of some installations may prohibit unsecured connections, so OPC UA Servers allow disabling the Discovery endpoint. When disabled, an out-of-band configuration must configure the Client with the connection endpoint of the Server. The two other endpoints support communications with IT systems using HTTPS and embedded devices using OPC UA Secure Conversation.

Server Types

Unlike other service sets in OPC UA, Discovery Services aren't always provided by the Server. In some applications, Servers can be used that only exist to provide Clients with Discovery information. In smaller applications, each Server is going to provide its own Discovery Services over a Discovery endpoint. There are "well-known" endpoints (Table 1) for each type of transport, so a Client knows where to get the Discovery information.

TRANSPORT	WELL-KNOWN ENDPOINT
HTTP	http://localhost/UADiscovery
OPC UA TCP	opc.tcp://localhost:4840/UADiscovery
HTTPS	https://localhost:4843/UADiscovery

Table 1 - Well-known Discovery endpoints

It's possible for several Servers to be located on the same platform or within the same local network. In that case, a specialized Server, called a Local Discovery Server (LDS), exists only to provide Discovery Services. The LDS can be located on the same host with a Client (LDS-ME) or with a Server or at a separate TCP/IP address.

Once a Server registers itself with an LDS Server, it turns over handling of Discovery messages to the LDS. Client devices can then access the LDS Server and get the list of Servers registered with it.

Servers send identifying information to the LDS Server in the Register Service message. That information includes the Server name, Server type, a product description string, the list of Discovery endpoints supported by the Server, and a short list of Server Capabilities.

The same Server registration is used with either standard LDS Servers or LDS Servers that use a multicast network. LDS Servers that operate on a multicast network are called LDS-ME Servers.

In very large systems, there are Servers that can track the location of Servers across an entire network, plant or corporate facility. These specialized Servers are known as Global Discovery Servers (GDS). A GDS differs from an LDS in that the address space of the GDS is designed to catalog a group of Servers. A Client can search the address space of the GDS to learn about Servers in the network.

Discovery Service Messages

The following tables provide a short introduction to the Discovery Service Set – the set of services that implement OPC UA Discover. For a more detailed description, refer to the OPC UA specification.

THE EVERYMAN'S GUIDE TO OPC UA

GET ENDPOINTS SERVICE

Directed To	The Discovery endpoint of a Discovery Server (an LDS or GDS) or a standard Server.
	transport[4]//hostname:4840/UAdiscovery
Request Parameters	An (optional) list of transports to filter the endpoints returned in the response.
Response Parameters	The list of endpoints in the Server (LDS, LDS-ME or Server) that support the transports listed in the request. If no transports are included in the request message, all transports of the Server are returned.
	The endpoint information includes an application description. The application description provides product information, the application name, its type (Server, Client, Discovery Server) and the list of Discovery endpoints in the device.
Description	The Get Endpoints service provides the requesting device with the endpoints supported by a Server. A Client uses the Get Endpoints service with a standard Server to identify the endpoint that best matches its application requirements. A Client uses the Get Endpoints service with a LDS or LDS-ME to identify the endpoint and Server that best matches its application requirements. A Server can use the Get Endpoints service to identify the best endpoint of an LDS to use for a Register Service request but typically it will use the well-known port that most LDS Servers support.

Table 2 - Get Endpoints description

---

[4] Transport refers to one of the supported transports. Three are supported at this time: OPCUA TCP, HTTP and HTTPS.

REGISTER SERVER SERVICE	
Directed To	The registration endpoint of a Local Discovery Server (LDS). An LDS always provides at least one endpoint that supports the Registration service. This service is only available over a secure communication channel. Normally the address for registration is going to be: transport[5]//hostname:4840/registration Sometimes, for security reasons, a user can configure the LDS to use a different endpoint. All Clients and Servers using that LDS Server would have to be know about that endpoint using some out-of-band configuration mechanism: transport[6]//hostname:4840/someendpointname
Request Parameters	The Server name, Server type, a product string, the list of Discovery endpoints supported by the Server and a short list of Server Capability features.
Response Parameters	The standard Response Header and no additional parameters.
Description	The Register Service Request Message is how a standard Server makes itself known to a Local Discovery Server.

Table 3 - Register Server Service Overview

---

[5] Transport refers to one of the supported transports. Three are supported at this time: OPCUA TCP, HTTP and HTTPS.
[6] Transport refers to one of the supported transports. Three are supported at this time: OPCUA TCP, HTTP and HTTPS.

FIND SERVERS SERVICE	
Directed To	The Discovery endpoint of a Local Discovery Server (LDS). transport[7]//hostname:4840/UAdiscovery
Request Parameters	A list (optional) of Servers that the Client is interested in finding.
Response Parameters	The Application Descriptions for the Servers that match the list provided in the request message.
Description	The Find Servers Request Message is the service a Client uses to obtain the list of Servers that have registered with the Discovery Server (LDS). The Servers can be limited to specific Servers specified in the request or all Servers registered with the LDS. The returned Application Description contains a product URI, the application name, the application type and the Discovery endpoints in that Server. The Find Servers request can also be sent to the Discovery endpoint of a non-LDS Server. The application Server returns its application description in the response message.

Table 4 - Find Servers Service Overview

Discovery Security

Because Discovery endpoints support limited functionality, no security is required. Any Client can request information on the endpoints supported by the Server.

One of the risks in the discovery process is that Clients may access a Discovery Server provided by some malicious entity and access a rogue Server. This can be mitigated by using Discovery Services with a secure transport like HTTPS (TLS/SSL). It can also be mitigated by validating the Server Certificate and ensuring that the Server contains a Server Certificate from a trusted source.

Another risk is a rogue Server registering with an LDS and being found by a Client who initiates a communication session. Clients can mitigate this by validating the Server Certificate and ensuring that the Server contains a Server Certificate from a trusted source. In practice, the LDS Server should not register a device over an unsecure connection so a Client should be able to trust the Servers registered with an LDS.

Another risk is using Multicast DNS (mDNS). Multicast DNS applications are, by their nature, insecure. Critical applications should either not use Multicast DNS Discovery Services or use OPC UA security and certificate trust lists to control access.

---

[7] Transport refers to one of the supported transports. Three are supported at this time: OPCUA TCP, HTTP and HTTPS.

Discovery Process – No LDS Server

With no LDS Server on the network, the Client must know the local host name (typically the TCP/IP address) of the Server. The host name is obtained through some out-of-band mechanism like a web page, front panel, or other mechanism.

In this architecture, a Client appends "UAdiscovery" to the local host name to form the Discovery endpoint. The Client may be configured to use a specific transport like OPC UA TCP or HTTPS. If not, it may progressively try contacting the Server using different transports until it finds a Discovery endpoint that the Server supports. The possible Discovery endpoints as of this writing include

https://localhostname:4840/UAdiscoveery
opcua://localhostname:4840/UAdiscovery
http://localhostname:4840/UAdiscoveery

More transports may be defined in the future.

Once a Discovery endpoint is identified, the Client issues a Get Endpoints message to discover all the endpoints supported by the Server. Optionally, the Client can specify the endpoints that match some specific transports. For example, the Client may only be interested in HTTPS endpoints. The Server would return endpoint descriptions for each HTTPS endpoint supported by the Server. From the one or more endpoints returned by the Server, the Client selects the endpoint that best matches its requirements and begins the connection process by issuing an Open Secure Channel request.

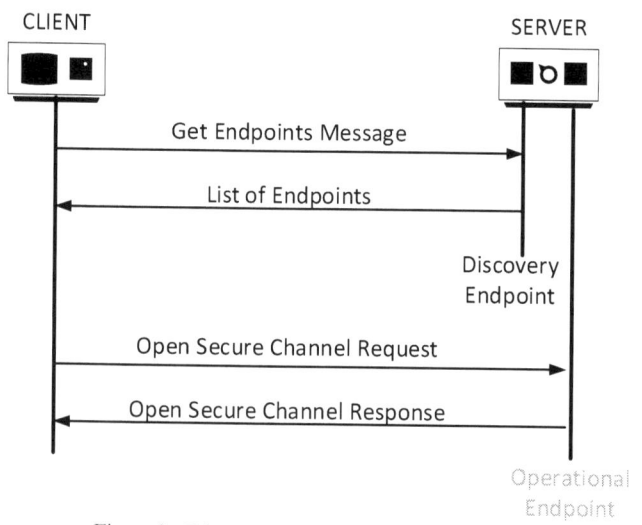

Figure 8 - Discovery process with no LDS Server

A Client may issue a Find Servers request on the Discovery endpoint of a non-LDS Server, but only the Application Description for the Server would be returned to the Client.

Discovery Process – With LDS Server

An LDS Server presents the same problem to the Client as the standalone Server. The Client has no sure way of finding the LDS Server. There are three ways a Client might find an LDS Server. One, get the specific address through some out-of-band mechanism like a web page, front panel, or other mechanism. Two, use the hostname of a Server to see if that host is also supporting an LDS Server. Or three, use its own hostname to see if there is an LDS Server active on the platform on which it's running.

Before the Client can request the list of Servers from a Local Discovery Server (LDS), Servers must register with it. Two services are available to the Server to accomplish this. First, the Server should use the Get Endpoints request to obtain the list of endpoints from the LDS. Once it chooses an endpoint with a compatible transport protocol and a sufficient security profile, it can open a Secure Channel and send the Register Service request message to get registered with it.

The Server includes its application description in the Register Service request. The application description includes its name, its type, a product string, the list of Discovery endpoints it supports, and a short list of Server Capability Identifiers. Server Capability Identifiers are short strings that map to OPC UA features. The identifiers include strings like "LD" (live data), "HD" (historical data), "RED" (redundancy) and "AC" (alarms and conditions). These identifiers assist the Client in choosing a Server that meets its application requirements.

The Client requests the Server information and Server Capability Identifiers using the Find Servers service on the unsecured Discovery endpoint of the LDS. The response to the Find Servers request is a set of Application Descriptions. A Client can request application descriptions for a specific set of Servers in the Find Servers service request.

Once the Client has application descriptions for Servers, it next issues Get Endpoint requests to identify endpoints in those Servers that meet its criteria. The Client can optionally request endpoints that meet one or more kinds of transports. The Server responds with endpoints for all its available transports, or only the transports that match the list of transports specified in the original request message.

If the Client finds an endpoint that meets its requirements for compatible transport, security policy and security mode, it can start the connection process by issuing an Open Secure Channel request.

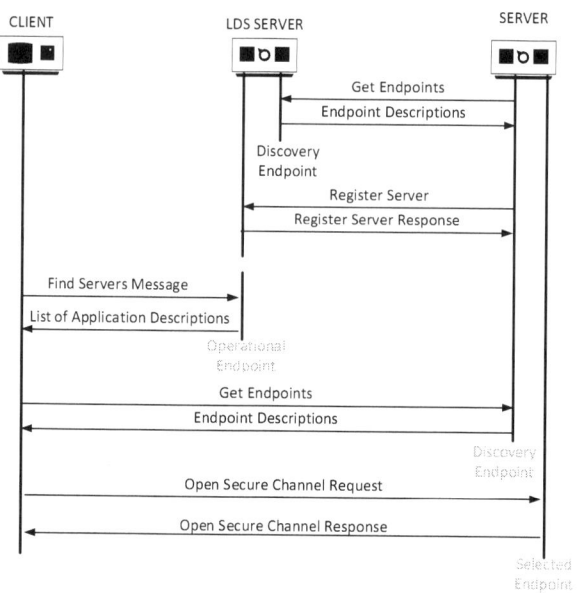

Figure 9 - Discovery process with LDS Server

Discovery endpoints are unsecured endpoints in application Servers and Discovery Servers, and the security policies of some installations may prohibit unsecured connections. OPC UA Servers allow disabling the Discovery Endpoint of a Server in these installations. Some out-of-band configurations must configure the Client with the connection endpoint of the Server.

Discovery Process – With LDS-ME Server

A Discovery Server that supports the Multicast Extension (LDS-ME) is a Discovery Server that uses a multicast network to identify OPC UA Servers. From networking basics, you'll remember that multicast networks using addresses in the range of 224.0.0.0 to 224.255.255.255 are networks where packets of information can be easily ignored if they are of no interest. If interested, a device can subscribe to the multicast network and receive those packets.

Discovery Servers with the Multicast Extension (LDS-ME) use the multicast network to both get information on Server devices and provide that information to interested Clients. The Server capability indicators, the same endpoint descriptions and the same application descriptions from the last section, are used by LDS-ME Servers and Clients when communicating over a multicast network. Only the way LDS-ME Servers communicate with Clients and other Servers is different when using multicast Discovery Servers.

LDS-ME is mostly used in an environment in which a great many sensors or other small UA Servers exist. These platforms do not provide anything but a simple UA Server (no Discovery Server). This allows any LDS-ME Server that is on the network to collect all of the multicast messages and identify all of the Servers on the network. Additional security configuration may be required in a network like this but that is outside of the scope of OPC UA.

For Servers, the registration process to register with either an LDS or an LDS-ME is identical. A Server registers itself when it comes online or goes offline. The Server follows the same process to determine the address of the Server as described in the previous section on LDS Servers without the Multicast Extension. The Server uses the same services to register with an LDS Server as with an LDS-ME Server.

For Clients, an extended version of the Find Servers service described earlier is available to work with the multicast network. Clients can use the same Find Servers service (described earlier) to get a list of Servers on the local network, or they can use the new "Find Servers On Network" to get the list of Servers that have registered with an LDS-ME on a multicast network.

Multicast DNS, or mDNS, is the multicast networking technology behind the LDS-ME Server implementation. mDNS provides the infrastructure for announcing the Servers on the network and communicating with remote Clients that want access to those Servers. There are many sources on the Internet to learn more about mDNS. mDNS is the same technology that is used in most home networks to detect network printers.

Unfortunately, mDNS is not considered a highly secure technology. Some installations prohibit it, but if you decide to use it with OPC UA, you should use the highest level of security offered by your OPC UA devices.

Like with LDS Servers, before the Client can request the list of Servers from a Local Discovery Server (LDS), Servers must register with it. Again, the Server should use the Get Endpoints request to obtain the list of endpoints from the LDS-ME Server. Once it chooses an endpoint with a compatible transport protocol and a sufficient security profile, it can open the secure connection and send the Register Service request message with all the parameters described earlier.

Other than using the Find Servers On Network service, Clients use a process identical to the one used for local LDS Servers. The process for Server Registration and Client access is illustrated in Figure 10.

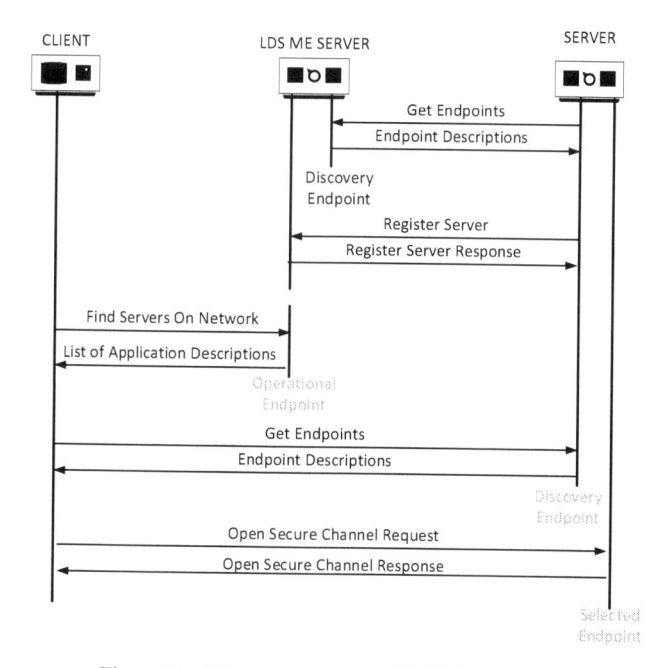

Figure 10 - Discovery process with LDS-ME Server

Discovery Process – With Global Discovery Server

A Global Discovery Server (GDS) is a device that is radically different that the Discovery Servers described previously in this chapter. A GDS is an OPC UA Server whose address space is a representation of the OPC UA Servers available in a system. Instead of asking for Server information like a Client might do with a LDS, devices can browse the address space to learn about what Servers are on the network.

As of the time of this writing, few GDS Servers are in service, and those are more beta test units than practical, production units. GDS Servers are usually coupled with certificate management services.

Hints for Reading the OPC UA Specification

Part 12, Discovery Services, is an excellent description of OPC UA Discovery. It is well-written, informative and a wonderful guide to how OPC UA Clients discovers Server devices. If you wish to thoroughly understand how OPC UA Discovery Services use multicast networks, you may wish to study Multicast DNS prior to studying Part 12.

You will get a much more thorough treatment of OPC UA Discovery Services in Part 12 for:

THE EVERYMAN'S GUIDE TO OPC UA

- The kinds of Multicast DNS network architectures that you might find in across the plant or across the corporation OPC UA networks.
- How specific commands referred to in this chapter are actually implemented. For example, Part 12 details the differences between the RegisterServer and RegisterServer2 services.
- The Information Model and its implementation in a GDS Server's address space.
- An Information Model for certificates and a mechanism for managing certificates.
- The complete list of Server Capability Identifiers.

ADDRESS SPACE

What's Interesting?

What's interesting about OPC UA is how the OPC UA address space is both simple and incredibly flexible. There is no other Industrial Automation technology that can make that claim.

The simplicity of the OPC UA address space model commences with its base element, the node. A node is simply a highly structured data element consisting of a set of predefined attributes and relationships (references) to other nodes (see Figure 11).

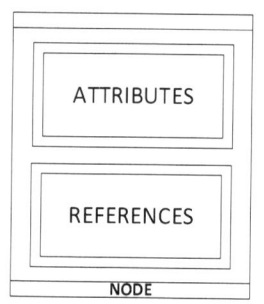

Figure 11 - The structure of an OPC UA Node

Using this very simple data element, an OPC UA address space can be created that describes very complicated processes. The flexibility of the address space allows a designer to present not only raw process data, but also extensive information about the state of the underlying process and the process environment. This flexibility ensures that any system, no matter how complicated, can be exposed using OPC UA.

What Do You Need to Know?

The objective of the OPC UA address space model is to standardize the way that automation devices represent processes, systems, and information to Client devices. Processes, systems and information are represented as Objects in OPC UA. Objects are collections of nodes, simple or complex, that together represent both the process state and the process environment.

Nodes

Nodes are the basic elements of the UA address space model. Nodes are categorized by a Class and composed of a set of attributes and one or more references to other nodes. Of the eight Node Classes defined in OPC UA, the two most important are the *Object NodeClass* and the *Variable Node Class*.

A node of the *Object NodeClass* is an organizing node that a designer uses to represent an entire real world machine, process, system, or some component of it. The root node of every UA address space is always a node of the *Object NodeClass*.

Variables of the *Variable NodeClass* are the only nodes in an OPC UA address space that contains data values. A node of the *Variable NodeClass* either contains a real world data value or a meta-data value that characterizes a node. A node with meta-data is known as a property and is referenced from the node it characterizes by a *hasProperty* reference. Properties characterize nodes with meta-data like engineering units, maximum or minimum value, last configuration time, or any other data the address space designer wishes to include.

Attributes

Every node of an OPC UA Node Class is described by one or more of the twenty-one predefined attributes. Some attributes used by all Node Classes including:

- Node Class – The node category or Class for the node
- Node Description – A Local Language description of the node
- Node Identifier – The unambiguous identifier for the node
- Node Display Name – The text string used to display the name of the node

References

References are essentially pointers that relate nodes to one another. A reference is of a specific type and identifies both the source node and the target node. Some of the most common reference types include the *hasTypeDefinition* type, which identifies the type definition; the *hasComponent* type, which implies that a node is, in some fashion, an element of another node; and the *hasProperty* type, which identifies that there is a node which contain some information characterizing the source node.

A snippet of an address space is presented in Figure 12. This address space snippet illustrates how a node can be a component of another node and how a variable can have a type definition and a property. The variable *OvenTemp* is a component of the Heat Sensor Object. It has a type definition which defines it, and a property reference to a maximum temperature. The *MaxTemp* property value characterizes the variable *OvenTemp* node by storing the maximum oven temperature.

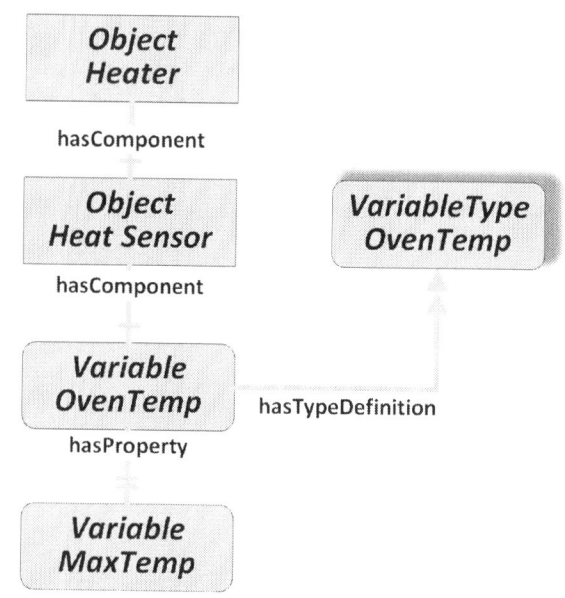

```
            Object
            Heater
              |
         hasComponent
              |
          Object              VariableType
        Heat Sensor             OvenTemp
              |                     ↑
         hasComponent               |
              |                     |
          Variable  ————— hasTypeDefinition
         OvenTemp
              |
         hasProperty
              |
          Variable
          MaxTemp
```

Figure 12 - Object & Variable Node Class example

What's the OPC UA Advantage?

The address spaces used by most automation technologies are flat. Modbus RTU and Modbus TCP are the best examples of a flat address space. Modbus has 64K of register space and 64K of coil (bit) space. There are no organizing elements that define any type of structure to all those registers and coils.

There are other technologies that use an Object structure, but there is no way to define a set of Objects to represent the structure of a system. A conveyor that contains two elevators both containing motors, drives, switches, and sensors cannot be represented in EtherNet/IP or ProfiNet IO in a way that communicates the structure of that system. The individual elements, their attributes and values, can be represented, but there is no mechanism for creating any higher level structure.

This is an important distinction, as in OPC UA, Client devices can browse through the structure of the system and identify the components that form the system and discover their relationships.

XML, even though it is not very useful for some automation applications, is a technology that has at least some ability to organize information. In XML, any number of nested child tags can be created to mimic the structure of a system. It also includes the concept of values and properties, like OPC UA. Like UA, it offers a version of typing where the Object structure can be reused. Where it fails is in the type of data that can be organized (everything is ASCII) and how difficult it is for resource-constrained devices to encode and decode XML.

What are the Details?

The OPC UA address space is the key to OPC UA. Everything else is useless without an address space. All of the transports, connections, and services exist to provide a Client with access to the address space. Understanding the address space model and how Objects are represented by nodes in the address space is an important key to understanding OPC UA.

If you remember chemistry class, each element in the periodic table is a substance that cannot be decomposed into other elements. In the OPC UA address space, that elemental structure is a node. It is the core of all the Objects, Types, references, and everything else that forms an OPC UA address space.

An OPC UA node is composed of attributes and references (Figure 1). Attributes are the primary characteristics of a node. OPC UA defines twenty-one attributes. The number of attributes is fixed. Address space designers can create no additional attributes. References connect nodes to one another and give the ability to create a hierarchical structures of nodes.

OPC UA nodes can be grouped into three general types: Data Type nodes, Instance Type nodes and instances of those types.

Data Type Nodes

Data Types nodes define a particular kind of data. Data Type nodes can define a common type like Integer. In many cases a Data Type node will define a type specific to an industry. In the oil industry, there is a variable called Rock Porosity. An OPC UA variable with a Rock Porosity data type would certainly have a reference to a Data Type node that would explain to a Client device how to interpret variable types by the Rock Porosity data type.

Instance Type Nodes

In a similar way, Instance Type nodes provide definitions for Objects, Variables, and Methods. A Drill Bit Object in the Oil and Gas Industry might have a Drill Bit Object Instance Type which identifies all the component nodes that comprise the Drill Bit Object.

Instance Nodes

Instances of a Type are the actual instantiation of the Object Type. The Drill Bit Object is the instantiation of the Drill Bit Object Type. These are the nodes that actually encapsulate Drill Bit data as opposed to the Instance Type nodes that define the nodes in the Drill Bit type.

Node Classes

Data Type nodes, Instance Type nodes, instance nodes, and all other UA nodes are categorized by a Node Class. The OPC UA specification defines eight Node Classes (see **Table 5**).

NODE CLASS	NODE CLASS IDENTIFIER	DESCRIPTION
OBJECT	1	Object nodes provide an abstract representation. An Object can represent an entire process, system, or machine or an individual component of the process, system, or machine.
VARIABLE	2	Variable nodes can be either Data Variables, which represent values or Properties, which characterize the values of other nodes.
METHOD	4	Method nodes represent lightweight functions that can be initiated on a Client request.
OBJECT TYPE	8	Object Type nodes define the structure of one or more Objects that represent a real world process, system, or machine.
VARIABLE TYPE	16	Variable Type nodes define the structure of Variables. A Variable Type is often used to define a variable used multiple times in an address space or in devices specific to a particular industry.
REFERENCE TYPE	32	Reference Type nodes define common types of references used in an address space. Just like variable types, a reference type is used when a reference is used multiple times in an address space or in devices specific to a particular industry.
DATA TYPE	64	Data Type nodes define a particular kind of data. All OPC UA devices are expected to know standard data types like Integer and Float. Trade Groups, Vendors and others can define specific data types common to their application areas.
VIEW	128	View nodes specify a subset of the nodes for those cases when Client devices only have an interest in a sub-section of the address space.

Table 5 - Node Class table

Node Attributes

Every node is described by one or more of the twenty-one predefined attributes that characterize the node. The nodes of each Node Class make use of some portion of the twenty-one attributes. Some nodes use more and some use fewer of the twenty-one attributes but all use the Node Class attribute, the node description attribute, the node identifier attribute and the Display Name attribute.

Nodes of a particular Node Class are required to implement the mandatory attributes for that Node Class. Table 6 lists the Mandatory and Optional attributes for each Node Class.

THE EVERYMAN'S GUIDE TO OPC UA

Attribute	Var	VarType	Obj	ObjType	RefType	DataType	Method	View
AccessLevel	M							
ArrayDimensions	O	O						
BrowseName	M	M	M	M	M	M	M	M
ContainsNoLoops								M
DataType	M	M						
Description	O	O	O	O	O	O	O	O
DisplayName	M	M	M	M	M	M	M	M
EventNotifier			M					M
Executable							M	
Historizing	M							
InverseName					O			
IsAbstract		M		M	M	M		
MinimumSamplingInterval	O							
nodeClass	M	M	M	M	M	M	M	M
NodeID	M	M	M	M	M	M	M	M
Symmetric					M			
UserAccessLevel	M							
UserExecutable							M	
UserWriteMask	O	O	O	O	O	O	O	O
Value	M	O						
ValueRank	M	M						
WriteMask	O	O	O	O	O	O	O	O

Table 6 - Mandatory and Optional Attributes

Node References

Nodes are connected to other nodes through references. References are essentially pointers that relate nodes to one another. A reference is of a specific type and identifies both the source node and the target node. Some of the most common reference types include

- *hasTypeDefinition* – connects a node to its type definition.
- *hasComponent* – connects a node to another node in a hierarchy. The implication of the *hasComponent* reference is that the target node is, in some fashion, a part of the source node.
- *hasProperty* – identifies nodes that contain some information which characterizes the source node. Properties can be anything including units, minimum value, maximum value, manufacturer name, last configuration date, or anything else the address space designer wants to include in the address space.

Address Space Example

Collections of nodes with their attributes and references represent real world systems. Figure 13 provides a partial address space for the collection of nodes that might represent the heater of a curing oven. The four nodes at the top of the diagram are OPC UA system nodes. These nodes (***Root Object, ViewFolder Object, ObjsFolder Object, ObjectTypes Object***) are found in every OPC UA Server. The nodes below those nodes describe the curing oven heater. (Note that for clarity, the component nodes for both the Conveyor Object and the Timer Object are excluded.)

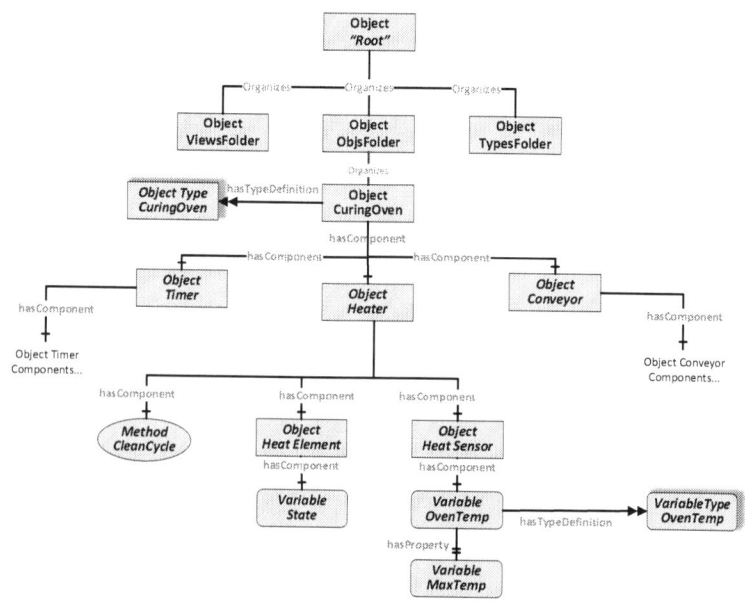

Figure 13 - Heater Object address space

If you closely examine Figure 13, you will note that there is a *hasTypeDefinition* reference to the CuringOven Object Type. That object type indicates that there is a definition for the CuringOven Object. Object definitions are templates for Object definitions. They provide both required and optional objects that will be found in an Object instance. An Object definition can be used to ensure that all systems in an application domain are using the same representation of a system, process or machine.

Object definitions are created by any one of several sources:

- The OPC UA Foundation – The Foundation makes available commonly used Object definitions that can be found in many applications.
- Trade associations – Various trade associations make Object definitions for common devices in their industry available to automation vendors in their industry.
- Vendors – Vendors can create their own definitions that describe their specific machine, process, or system.

These definitions are available, much like XML schemas are available over the Internet. In some cases, a product vendor may include the definition of the Object Type right in the product address space. In that case, a Client can follow the *hasTypeDefinition* reference and obtain the structure of the Object.

A Note On Creating Address Spaces

This section is for vendors creating OPC UA devices. At the time of this writing, creating an address space is a somewhat unwieldy task. This process will improve as time goes on. The current steps required to create an address space are shown in Figure 14.

STEP 1: CREATE DATA MODEL OF YOUR DATA

The first step in creating an OPC UA Application Data space is to convert the data you want in the address space into a Data Model. OPC UA has a specific format for this Data Model XML file.

Depending on the number of unique object and data types, this process can be a simple manual process or a process that requires one of the OPC UA Data Modeling Tools.

STEP 2: CREATE NODESET XML FILE

The next step is to convert your data into a list of nodes that will comprise your address space.

OPC UA provides a complier to create the user Nodeset from a data model XML file.

STEP 3: ADD STANDARD UA NODES

The user Nodeset file does not contain the standard UA Nodes that have to be in every UA Application space. This step merges the UA Nodeset file with the User Nodeset file to create a complete Nodeset

STEP 4: CREATE ADDRESS SPACE FILE

Once the Nodeset is complete the Address Space can be created. The Address space file is a standard ANSI C file that is compiled into your project and provides the definition of the Address Space to the OPC UA Source Code stack.

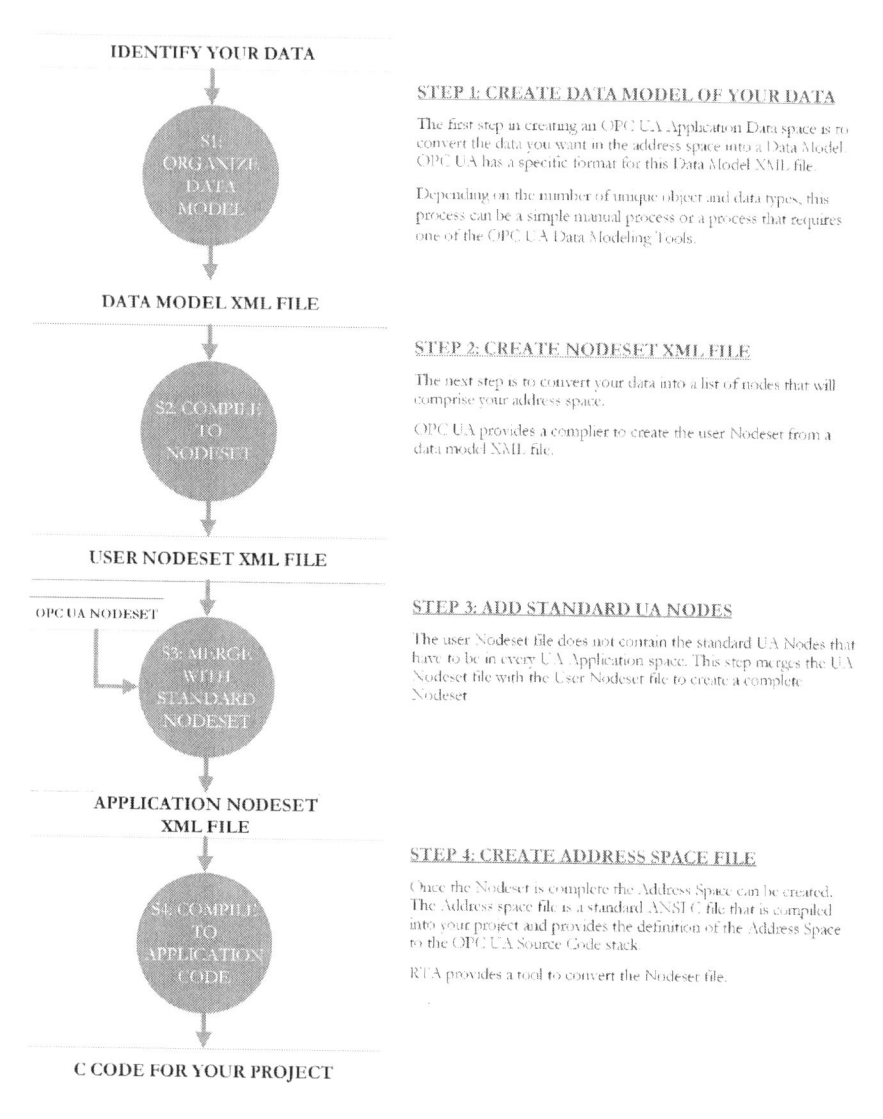

RTA provides a tool to convert the Nodeset file.

Figure 14- Address space creation process

Hints for Reading the OPC UA Specification

Part 3, address space reference, is one of the most easily understandable components of the OPC UA specification. Sections 3, 4, and 5 are excellent descriptions of the address space model and wonderfully describe each UA Node Class.

THE EVERYMAN'S GUIDE TO OPC UA

Sections 6 and on are more complex, less understandable, and not particularly relevant except to OPC UA device designers that really wish to use the typing system defined in OPC UA.

INFORMATION MODELING

What's Interesting?

In lots of ways when you compare the technology used in Industrial Automation (IA) to what's happened to IT over the last five, ten, or twenty years, IA has been pretty unsophisticated for a very long time. The reasons for that are well understood and they're good reasons: large capital costs for equipment, long term operational requirements, the costs incurred to upgrade technology, and all the rest.

But that's changing now. It has changed a lot in the last several years and it's going to change more. There are new requirements now, and to meet those new requirements one of the areas that's going to have to change is how we model information. We can't meet the new requirements for forward integration with our customers, backward integration with our vendors, and integration with our enterprise systems without better mechanisms for modeling information.

What's interesting about Information Modeling in OPC UA is that OPC UA is the only technology that has standardized the documentation, implementation, reference, and access to Information Models. OPC UA allows end users, trade associations, vendors, and others to create, distribute and implement Information Models in a very organized, detailed, and structured way.

For example, trade associations like the Undersea Oil and Gas Association (MDIS), the Building Automation Industry (BACnet), and the RFID Association all have recognized the need to standardized access to devices in their industry, distribute those standardized models in a cohesive way, implement those models in a structured way, and reference those models in such a way that devices know exactly what information is available and how to access it. That's very powerful and it's why these trade associations and others have standardized on OPC UA.

What Do You Need to Know?

What is an Information Model? An Information Model is nothing more than a logical representation applied to a physical process. An Information Model can represent something as tiny as a screw, a component of a process like a pump, or something as complex and large as an entire filling machine. The Information Model is simply a structure that defines the component, devoid of any information on how process variables or metadata within that structure can be accessed.

Figure 15 is the Information Model for a pump. It details all the significant components of the pump and the information that is available within the pump. Other than the structure implied by the way the information is related, there is no detail on how the data is stored or how it is accessed.

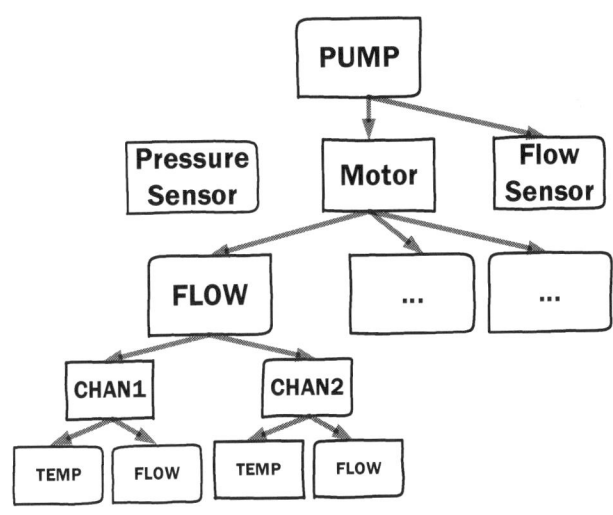

Figure 15 - A snippet of a pump Information Model

And that last part is the important point. The Information Model has nothing to do with how that information is stored or made available. In OPC UA, Information Models are XML files – highly structured files that can represent anything from a variable to a device to a machine, or even a plant. That XML file is implemented in the OPC UA address space and there are specific mechanisms to browse that address space, access the nodes that comprise it, and encode, secure, and transport the information it contains. All of that is important, but in OPC UA those mechanisms are distinct and separate from the how the information is modeled.

It's important to understand that OPC UA is the first technology to separate this functionality. Other technologies exist to model information. Other technologies exist to represent data and information in an electronic system. Very few if any, provide both while still maintaining a distinct separation between the two like OPC UA does.

What Are the Details?

The factory floor is a place where none of us expected to have to process massive amounts of data. But that is exactly where we now find ourselves. Not only do our factory floor systems collect more data than we ever expected, we have requirements to move that data to unexpected destinations. Machine status must be moved to the cell phones of control engineers. Diagnostic data to the maintenance organization's Servers. Production data to the enterprise business systems and applications. And archived quality data to the cloud. We're swimming in so much data and places to send it that it's almost ludicrous. And unfortunately there is no reason to believe it won't get worse in the future.

Capturing all that data, securing it, moving it, storing it, and analyzing it: that's now a reality for those of us architecting manufacturing systems. And we can't accomplish that effectively without better mechanisms for organizing all that data. The flat files systems that factory floor controllers and systems used in the past just aren't adequate for this explosion in data. It's time we adopted the technologies and practices of the people that have being doing it for all these years: our friends over in the IT department. Organizing all that data as an Information Model is the difference between swimming in all that data and drowning in it.

What Is an Information Model?

But what is an Information Model? An Information Model is a representation of concepts, relationships, constraints, rules, and operations that capture the characteristics of an entity or process. It provides a sharable, stable, and organized structure for communicating information about that system or entity. The Information Model is simply a structure that defines the component, devoid of any information on how those characteristics captured by the model can be stored or accessed. There are languages designed specifically for Information Modeling like the Unified Modeling Language (UML). Figure 16 illustrates a portion of the Information Model for Wikipedia.

THE EVERYMAN'S GUIDE TO OPC UA

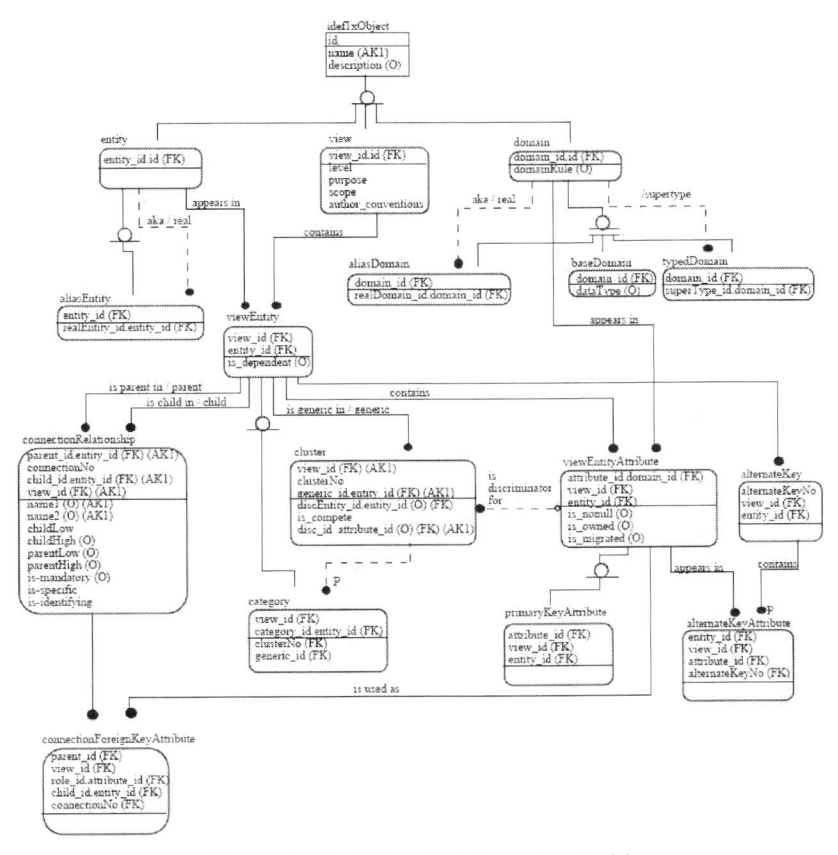

Figure 16 - The Wikipedia Information Model

The first thing you do when creating an Information Model is to decide what is of interest and what isn't. The Information Model for a centrifugal pump might only contain the color, case style, manufacturer part number, and purchase data if your Information Model is designed for asset tracking. It might contain none of those items if your Information Model is exclusively used in your manufacturing operations. As part of your process information it might contain the current RPMs, operating hours, and gallons per minute. Or, to support a number of different applications, it might contain both.

Once you pick those items of interest (entities in database lingo) and you define their specific characteristics, you set up the relationships between those items. A filling machine has all kinds of devices: valves, pumps, motors, controllers, and sensors which we can call Objects in the model. Each of these Objects are modeled by more specialized Objects. A Motor Drive Object might consist of a Motor Object and a Drive Object, for example. And those Objects can be modeled by other, more specialized

Objects. As you do this, a hierarchy of Objects is developed that forms the Information Model for the system.

The hierarchy of Objects for a curing oven is shown in Figure 17. This model is composed of three component Objects: a Timer, a Heater, and a Conveyor. The hierarchy of component Objects for the Heater Object is also shown. Note that the entire hierarchical structure of the curing oven is encapsulated in its type, the Curing Oven *objectType*. This Object model can be reused by instantiating another instance of that *objectType*.

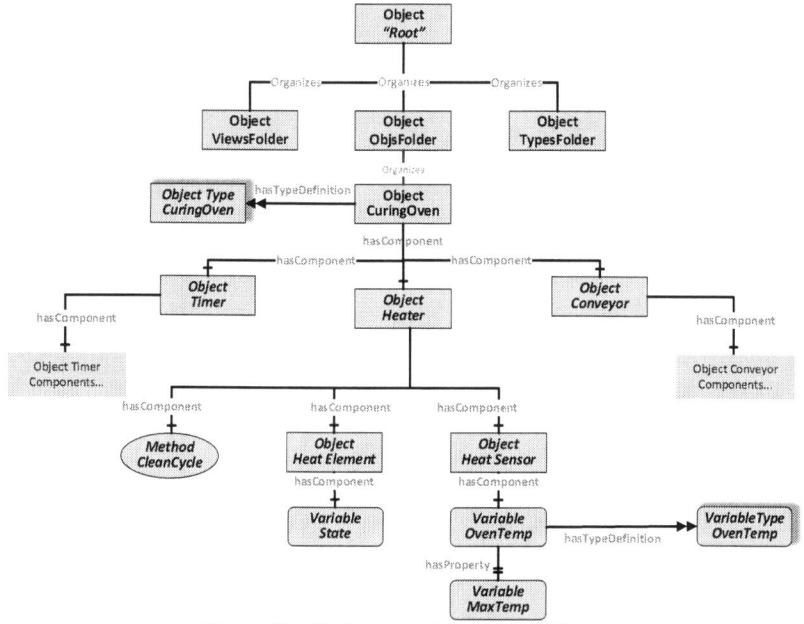

Figure 17 - Curing oven Information Model

You can get as complex or as straightforward as you'd like with an Information Model. It has infinite flexibility to describe your process in whatever way serves you best. When complete, you can document your process using a standard language and symbology that conveys to everyone exactly what each entity is and what relationship exists between those entities.

But what have we really done? Your Information Model is only that – your Information Model. You may have modeled a pump with the characteristics speed and RPM. Someone else might have used a pump model that includes the current flow rate. The RPMs might be an integer value in one system and a float in another. Since you've both modeled the pump differently there is no saving in labor or productivity for any of your customers. You may have given them a model using some open standard but they still have to incorporate your proprietary characterizations of the pump. That leaves us where we've always been: repeating the integration again if we have to use a different pump in the next application.

It's actually worse than that. We haven't even begun to talk about common transports, data encoding, and access to the data contained in the Information Model. It's one thing to define a nice Information Model for your device, your machine or production line but if there isn't any standardized way for others to know that you are using that model, to know what's in it, and to easily access it, it doesn't save anybody any time or money.

What's Different About UA Information Modelling?

And that's the problem that many of us have faced over the years. Yes, there have been people that have created very elegant Information Models and there are wonderful ways of documenting those models. In fact, there are languages, like UML, dedicated to Information Modeling. But the big problem in the past was that a lot of the models were tightly integrated with a specific technology or communication protocol. In other cases, the Information Model was designed specifically to solve problems in one particular application domain. And when there was a mechanism to create a flexible Information Model, there wasn't a standardized way to deploy the model in an actual system.

OPC UA solves these problems. OPC UA is the first technology to provide system architects with a common infrastructure for modeling data and providing the transport, encoding, and security to employ it. Information Modeling in OPC UA provides enhanced organization, flexibility, and scalability. It allows you to customize your view of your information to agree with the way you do business and how you integrate with other enterprise applications, improve asset performance and reliability, and act on your available data with greater agility and confidence.

OPC UA Information Modeling is different than other technologies in many ways:

- There is a consistent structure and standardized definitions that are flexible and adaptable for many application domains.
- There is a consistent and standardized structure to the documentation of the model (XML).
- There is a mechanism for translating the Information Model into a real time data model (also called an address space).
- There is a standard mechanism for Clients to identify the model and access component definitions and type information at run time.
- The encoding, securing, and transporting of values in the real time address space are entirely disconnected from the Information Model development.

The next few sections describe how this all works in practice, from specifying the model to deployment.

How Is an OPC UA Information Model Specified?

OPC UA Information Models use the concept of an Object to represent real world entities. An OPC UA Object can represent a simple device like a photo eye or even just the LED of the photo eye. Or it could represent a manufacturing cell, a production line, or even an entire plant.

Objects are formed from component Objects, variables, methods, and references to other Objects. Variables represent the static and variable data values that are contained in the Object. Methods represent small logic segments that can be executed by a Client device. While variables and methods are very important to OPC UA Information Modeling, references provide the glue that binds the components of an Information Model together.

OPC UA provides a rich set of references: *Organizes*, *hasTypeDefinition*, *hasComponent*, *hasProperty*, and many others. In particular, the *hasComponent* and *Organizes* references provide the ability for a designer to create hierarchical structures that mimic the hierarchical structures found in all real world systems. A *hasComponent* reference signifies that one entity is an integral and important element of another entity. An *Organizes* reference signifies that a node is simply part of a group of nodes that are collected together for some purpose. Figure 17 from earlier in this chapter provides a snippet of an Information Model for a curing oven. The hierarchical nature of the Information Model is apparent in this diagram.

While *hasComponent* references provide for hierarchical relationships, the *hasTypeDefinition* references provide the real power in OPC UA Information Modeling. At its simplest, the *hasTypeDefinition* simply points to the Object definition. That definition could be as simple as indicating that every instantiation of the type includes a variable. In much more complicated definitions, the *hasTypeDefinition* can point to a definition which includes a large number of Objects, variables, references, and methods.

There is no limit to how many Objects can be instantiated from a single *objectType* definition. The *objectTypes* pointed to by the *hasTypeDefinition* reference can be instantiated as many times as the entity it represents exists in the system, with every instance identical to all other instances[8].

For type definitions common to multiple installations or throughout an industry, the type definitions can be located remotely to the Server where they are instantiated. In these cases, the type definition reference includes a common URI (Uniform Resource Identifier) used for all definitions of the type.

This is particularly important to trade associations, which understand

---

[8] In actuality the specification provides for optional objects and some degree of customizability.

that integration costs can be dramatically lowered if everyone uses a standard definition for entities specific to an industry. This kind of extensible Object typing, Data typing, and Object Modeling are some of the reasons why the Oil and Gas Industry, along with major vendors like Emerson, Honeywell, Invensys, and others, are standardizing on OPC UA.

Supportable and adaptable standard interfaces to common entities are critical to implementation of these systems. The devices themselves are well understood and solidly engineered. The majority of system problems are in the interfacing of devices between vendors – sometimes competing vendors. And unfortunately interface problems are often discovered at the worst possible time – deployment and startup in a distant and hazardous location – when the problems have the most impact on schedule and cost the most to resolve. The OPC UA Information Modeling standard goes a long way towards solving this problem.

How Is An OPC UA Information Model Presented?

The OPC UA Foundation has created a standardized graphical notation for presenting Information Models. Both the graphic Objects and the notation used to identify references from node to node are standardized.

In the standardized notation in the OPC UA specification, the Object shape is used to identify the Object Class. Rectangles identify Objects of the Object Class. Shadowed rectangles identify Objects of the *ObjectType* Class. Variables of the Variable Class are represented by rectangles with rounded corners. Variable types of the *variableType* Class are represented by shadowed rectangles with rounded corners. Each Object Class is assigned a specific shape and shadow.

In a similar way, the line style between two Objects identifies the type of reference. References are always noted by a line beginning at the source of the reference and ending at the target of the reference. A line ending in double solid arrows identifies a type definition. Lines containing a single vertical mark near the target of the reference indicate a component reference. Lines with double vertical marks identify properties[9].

The OPC UA graphical notation standardizes the graphic descriptions of Information Models across all OPC UA applications. Figures 3 and 4 illustrate the graphical notation of an OPC UA Information Model.

OPC UA also provides a non-graphical method of specifying an Object model, XML (eXtensible Markup Language). The OPC UA specification details a schema that can be used to document an Information Model. Clients that access a Server with reference to a non-resident Information Model have the option of finding the schema on the Internet, though it's not clear how many Clients will support that mechanism.

---

[9] See Part 3 Address Space for the entire set of graphical notations used in OPC UA.

How Is an OPC UA Information Model Implemented?

An Information Model isn't of much use until it is digitized and implemented to physically represent the entities it is designed to represent in an OPC UA Server. It might be nice to look at, but a Client device needs to be able to browse the Objects and access the real world data values modeled by the Information Model using OPC UA services.

The process to get from an Information Model to the data space in a Server where it can be useful varies with the vendor of the device. There are three general steps to this process:

1. Convert the model to a *Nodeset* – A model is little more than a nice drawing. The Objects, references, and variables of that model must be converted into a series of nodes, the base element of an OPC UA address space. The attributes of that node must be defined, node identifiers must be assigned, and the references from source to target nodes generated.

2. Add standard nodes – Models usually won't contain the standard nodes provided by the OPC UA Foundation and defined in the specification. These are the nodes that are part of every OPC UA Server and if they are not in the model, these nodes must be added to the *Nodeset* generated from the original Information Model. Figure 18 illustrates the curing oven example with a portion of the standard *Nodeset* included in the Information Model.

3. Make the *Nodeset* downloadable – The final step is to take the *Nodeset* and convert it to a form compatible with the Server software. This is the step that is the most vendor specific.

THE EVERYMAN'S GUIDE TO OPC UA

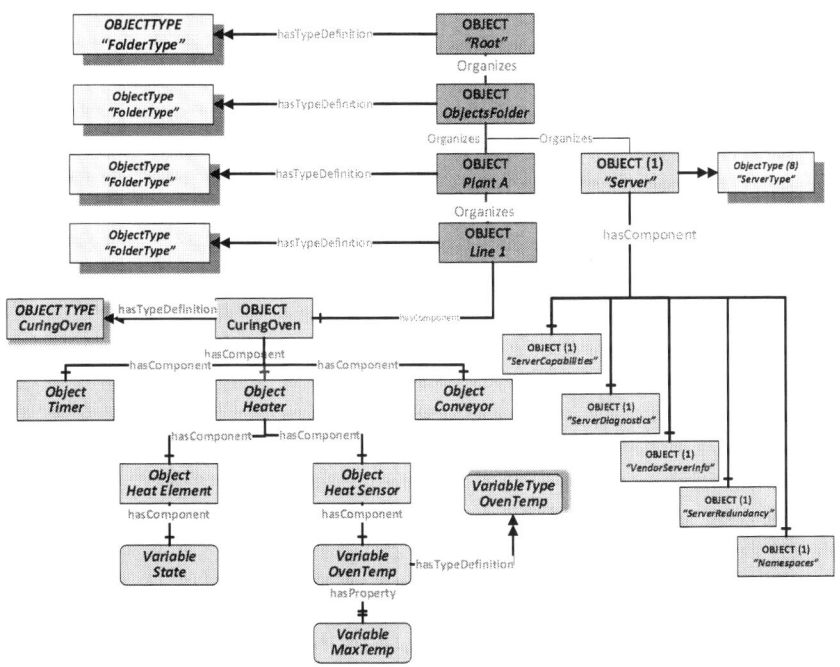

Figure 18 - Curing oven Information Model with standard nodes

How does an Information Model differ from an address space? You may note from the previous discussion in this chapter that the organization of the Information Model is identical to the address space organization discussed in a previous chapter. And you'd be correct if you noticed that. At first glance, there appears to be no difference between the OPC UA Information Model and the OPC UA address space model.

But they are different. The Information Model specifies the high level organization of the entity. It describes what the different parts are and how they are interrelated. The address space organization describes the specifics of how that model is deployed in an OPC UA Server device.

The Information Model exists to model the operation of a system, process, or entity. The address space model exists to provide the implementation details. The Information Model operates on a conceptual level. The address space model operates at the implementation level. For consistency and ease of implementation both use the same organization, based on nodes with attributes that characterize the node and references that link nodes from one to another.

How Are Information Models Referenced?

An OPC UA Client can use an Information Model in a number of ways. At its simplest, the Client can ignore the model and simply browse the

Objects of the model by following the *hasComponent* references in the address space. In Figure 18, starting at the root node, a Client can browse through the address space to find the Objects related to the curing oven.

This approach can lead to problems. If, for example, an item like the Oven Temp variable uses a structure defined in the model by the Curing Oven Trade Association, the Client will not know how to interpret it.

There are two possibilities: (1) the Client can browse to the *objectType* for the curing oven and look for the definition there; or (2) it can look for some external resource that contains the definition of the object type.

In the case of an external definition, the Oven Temp *variableType* is an OPC UA Object and like every Object in OPC UA, it is identified by a data type called a *NodeID*. One of the components of the *NodeID* is a name space index. The name space index is a URI (Uniform Resource Identifier) which identifies the naming organization responsible for assigning the *identifier* type and value. That way a group of Objects, like those of the curing oven, can be predefined by a 3rd party organization and Clients can know by the URI which organization is providing that definition.

Once the Client device notes that this Object or variable is defined by a name space index, the Client can either search the Server for the identification or find the definition externally. For vendor defined name spaces, it is likely that a Client can locate definitions by following the *hasTypeDefinition* references. For trade association definitions, it is more likely that the Client will have to download the XML Schema for the model to understand specific Object and variable definitions.

How Are Information Models Used?

Once an Information Model is digitized and represented as an address space, OPC UA provides the mechanisms to access that address space. OPC UA defines the service messages that Clients can use to read or write values in the address space. It provides a choice of mechanisms for digitizing or encoding those messages. It provides a set of security protocols that can be used to secure those messages. And it provides a selection of transport protocols that can be used to reliably move those messages between Clients and Servers. And all of this is independent of the definition of the original Information Model.

Hints for Reading the OPC UA Specification

Part 3, the specification for the address space, and Part 5, the specification for OPC UA Information Modeling, are the two parts of the specification that are pertinent to Information Modeling. There is a lot of common information in the two specifications but, in general, Part 3 is more of a basic primer on Node Classes and the structure of an address space (and Information Modeling, which uses the same basic structure).

Part 5 is more concerned with Object typing. It discusses the various kinds of *objectTypes* and *variableTypes*. Part 5 details all the kinds of types that are available. It details not only *objectTypes* and *variableTypes* but describes reference types, data types, standard methods, and standard views. You will especially want to read Section 8.3 on the Server Object to get a full understanding of this important Object.

You will want to pay special attention to Section 8.2 of Part 5. That's where the basic organization of an address space is described. It describes the root node and the concept of a folder in an OPC UA address space. That kind of detailed information would seem likely to be found in Part 3, but it actually is in Part 5.

DATA TYPING

What's Interesting?

One of those things that have caused a lot of gray hair for programmers over the years is variables that have a value of Null.

What is Null? Null communicates that the value of a variable is missing or unknown. That's pretty important a lot of the time. Knowing that you don't have a reliable data value to process allows you to make different decisions than when the value is zero.

Unfortunately, there hasn't always been a good definition of Null. Null in the C programming language is the character '\0' which, interesting enough, is not always the same as binary zero. That's a software bug that's probably bitten the backside of programmers something like tens of thousands of times.

OPC UA has fixed this for the variables where it matters most: string types, the *NodeID* node identification structure, all variable types based on the base data type, and most other variables. That's one less headache for a software engineer implementing OPC UA.

The other interesting feature of OPC UA is not that it provides the ability for trade associations, vendors, and others to create their own data types. It's that a Client can find the definition for these unique type definitions dynamically. That's something that is especially distinctive.

What Do You Need to Know?

Clients can't process data when they don't know how the data is formatted. When an OPC UA Client device can identify the type of data in a data variable, it can process it, manipulate it, and properly store it. Data typing is an important feature of all networking technologies.

OPC UA data typing is characterized by these important features:

1. OPC UA is a strongly-typed technology. Strongly-typed in this context means that variables have well-defined types and one type cannot be used in place of another type.
2. There is a predefined set of standard data types that all Clients and Servers must inherently understand.
3. Data variables are typed as one of these standard data types or as a type derived from one or more of the standard types.
4. The composition of types unknown to a Client are discoverable either locally in the Server or remotely using a URI (Uniform Resource Identifier).
5. A Client device can identify and match types from different Server devices as being identical types with the same origins and derivations.
6. The OPC UA typing system is extensible and new types can be created by vendors, trade groups, and other third parties to meet the requirements of a particular application domain.

OPC UA supports the standard data types found in many other technologies or computer languages. That standard type list consists of types you'll recognize like 16-bit Unsigned Integer, Double, Float and all the other ones you would expect to find in a standard type list.

The OPC UA Foundation has added a number of special data types to the architecture of OPC UA. For example, *NodeID*, a type defined by the Foundation, is on the well-known type list. *NodeID* specifies the unique identification of an OPC UA node.

Because well-known data types are inherently understood by all UA devices, the type definitions for these types don't exist in the address space. But other data types – creations of a particular vendor, trade association, or other 3rd party organization – must be available to Client devices that don't inherently recognize those types. Definitions for these types can be found either in the Server's address space itself or in some external location.

When new, unrecognized data types exist in the address space, they are organized as a subtype of the *BaseDataType* Node. This node is the root node for all data type definitions in OPC UA. Every data type that is not well-known is either a direct subtype of the *BaseDataType* or some derivation of one its subtypes.

A vendor can define specific data types related to their equipment or process that simplifies or enhances the operating, reporting, and diagnosis of that system. A vendor defines that type using some combination of base data types. They make their type known by publishing an XML definition of the type and creating a namespace URI (Uniform Resource Identifier) that specifically identifies the type (or group of types). A Client device without prior understanding of the type can search for it in the Server and, if not found, can sometimes find it externally if the vendor makes it available for public access.

What's the OPC UA Advantage?

All networking technologies type data. Some, like Modbus, have minimal typing and force all data into one of those limited types. Others offer a wider set of types that are based on a particular controller technology. A network that works with Rockwell or Siemens PLCs, for example, only exposes data types that align with those of that controller. OPC UA provides a more extensive and extensible data type system than any other networking technology.

What Are the Details?

Data typing and the terms "strongly-typed" and "weakly typed" are hotly debated by the geeks in the professional software community. In that environment, data typing refers to how much restriction a computer language like C, C#, PHP, Python, Java and all the rest place on what a programmer can do with a piece of data of a certain type. Language with weak typing or no typing at all will let a programmer do silly things like concatenate a string variable to a floating point variable. Languages with strong typing will detect these kinds of misuses and produce errors or warning messages when the source code is compiled.

For the purposes of the discussion in this book, we're going to misuse this term and classify OPC UA as strongly-typed even though OPC UA is an architecture not a computer language. In this context, strongly-typed for OPC UA means that all data is required to be typed and the definitions of types are either well-known or derived from well-known types.

Well-Known Types

OPC UA contains a rich set of these well-known, standard data types. Many of these types are standard data types found in many other technologies or computer languages. That standard type list consists of types like 16-bit Unsigned Integer, Double, Float, and all the other ones you would expect to find in a standard type list. All UA Clients and Servers are required to inherently understand the list of well-known data types.

Some of the common data types defined by in the OPC UA specification are listed in Table 7.

Byte	Common definition
Int16	Common definition
UInt16	Common definition
Int32	Common definition
UInt32	Common definition
Int64	Common definition
UInt64	Common definition
Float	Common definition
Double	Common definition
String	A length in bytes followed by a sequence of UTF8 characters without a Null terminator
DateTime	A 64 bit signed integer representing the number of 100 nanosecond intervals since January 1, 1601 (UTC)

Table 7 - OPC UA common data types

In addition to these well-known types that are common to many software systems, OPC UA Clients and Servers are also expected to know a set of types that are specific to the operation of OPC UA. These well-known types are listed here and in the tables later in this chapter:

- *Guid*
- *ByteString*
- *XmlElement*
- *NodeID*
- *Expanded NodeID*
- *StatusCode*
- *QualifiedName*
- *LocalizedText*
- *ExtensionObject*
- *DataValue*
- *DiagnosticInfo*

Because well-known data types are inherently understood by all UA devices, the type definitions for these types don't exist in the address space. The definitions of these types are expected to be implicitly understood by all OPC UA devices.

Derived Types

In addition to the standard data types you would expect, OPC UA provides for the ability to create or derive new types from standard data types. These "complex" types are types composed of one or more primitive types. **NodeID** is such a derived type. It is a structure composed of an Unsigned Integer, an enumeration and a value whose type is defined by the enumeration.

Vendors, trade associations, or other 3rd party organizations can create these derived types to meet specific requirements of their application domain. Derived types are of particular interest to trade groups – for example of the Pharmaceutical, Oil and Gas, or Barcode industries. These groups can make it much simpler to standardize access to common data in an industry by defining the data types for data in their problem domain.

Since derived types are unusable unless the Client device understands the derivation, OPC UA has two mechanisms for making those definitions known.

First, a Server can provide a node of the *DataType* Node Class to define the derived data type. A Client can access this node in one of two ways. The simplest way is for the Client to follow the *hasTypeDefinition* reference from the node in question to the type definition node.

The second way is for the Client to start at the root node and follow references to the Object *typesFolder*. All derived types with definitions in a Server will be a subtype of the *BaseDataType* Node (*DataType* Node Class) which can be found as an o*rganizes* reference from the *typesfolder*. This node is the root node for all data type definitions in OPC UA. Every data type that is not well-known is either a direct subtype of the *BaseDataType* or some derivation of one its subtypes.

It is not required that the definitions for derived types exist in the Server's address space. Type definitions can exist externally to a Server as they most likely would for the type definitions common to an industry and provided by a trade association. Type definitions for these kinds of types are going to be available in XML Schema files. These files define these application specific or vendor specific types in terms of well-known types just like all other derived types.

OPC UA places no restrictions on where these files can be located and does not define any operational processes for locating and accessing them. What it does is to create a mechanism for labeling a set of types as being defined by the same organization. The namespace element of the type identification indicates the origin of the type. It is identical for common, related types such as all those from a particular vendor or trade association.

In practice the namespace for a set of derived types could be something simple as "Hat Pin Trade Association". Or it might be a URI. The URI may be a placeholder like www.hatpinassoc.com/opcua/schema where that URL doesn't really exist. Or it may in fact be an actual URL where the type definition can be found. OPC UA does not define the contents of the namespace element used to uniquely identify a set of related types.

Type Identification

The namespace is part of the *NodeID* identifier. All types, well-known and derived, are identified by the built-in data type *NodeID*. The *NodeID* identifier identifies every node in a UA address space including the type definition nodes. Both well-known nodes and non-well-known nodes are identified by their *NodeID* value. A *NodeID* is actually a structure with three components as shown in Table 8.

namespaceIndex	A index to a table of namespaces
identifierType	An enumeration specifying the type of the next component of the structure: NUMERIC_0 STRING_1 GUID_2 OPAQUE_3
identifier	The actual identifier

Table 8 - *NodeID* Type structure

The *namespaceIndex* is an integer index into the table of all namespaces in the Server. The namespace identifies the responsible authority for assigning the identifier for the type. Namespace index zero is always the OPC UA Foundation and the name space table specifies the URI "http://opcfoundation.org/UA/". All well-known OPC UA data types are specified with a name space index of zero.

The *identifierType* field specifies how to understand the identifier. In its simplest form, the *identifierType* contains the enumeration "NUMERIC_0" which identifies the *identifier* field as having a numeric value.

The identifier field identifies the node in the context of the *namespaceIndex* and *identifierType*. The data type Int32, for example, is identified by *namespaceIndex* of zero (OPC UA Foundation), *identifierType* NUMERIC_0, and an identifier equal to six. Int32 is the seventh data type definition in the definitions made by the UA Foundation.

This type of organizational structure makes the definition of vendor or application specific data types open and available to anyone who wishes to use it. Vendors can even build Client devices that can automatically find definitions for devices they encounter on the Internet.

The *NodeIDs* for every OPC UA Foundation data type are published in the OPC UA specification and a file containing all the Foundation defined *NodeIDs* is available on the Foundation website.

Type Encodings

OPC UA variables and properties with these data types have to be transmitted from a Server to a Client at some point. To do that, the variable or property must be encoded as a byte stream. OPC UA currently supports two kinds of data encodings: binary and XML. The next chapter discusses OPC UA data encodings in detail.

Some Important OPC UA Derived Types

NodeID	
Description:	The *NodeID* uniquely identifies every node in the Server address space.
Primitive Type:	Structure of: • *nameSpaceIndex* – Index into the name space table in the Server Object where the URI for the naming authority of a *NodeID* is located. Index zero of that list refers to the OPC UA Foundation for the nodes defined by the Foundation. • *identifierType* – An enumeration specifying the type of the next component of the structure (NUMERIC_0, STRING_1, GUID_2, OPAQUE_3).
How It's Used:	Every node of every Node Class in the Server address space is identified by a *NodeID*. For example: *NodeID* {0, NUMERIC_0, 2007} refers to the node of the Variable Node Class with the name *ServerType ServerStatus*. *NodeID* {0, NUMERIC_0, 84} refers to the Object variable with the name *RootFolder*. *NodeID* {5, NUMERIC_0, 54} refers to an Object with the name *mySampleValveState* defined by the naming authority listed in the sixth entry in the namespace table.
Why it's important:	The *NodeID* is the identifier for everything in the Server address space. It classifies every node and provides the means for a Client to identify the origin of a node. Knowing the naming authority that created the node means that the Client can identify that two nodes in two different Servers refer to the same node, for example if a speed variable in a motor drive is the same as a speed variable in another motor drive.

Table 9 - NodeID Derived Type

THE EVERYMAN'S GUIDE TO OPC UA

Extension Object	
Description:	The Extension Object is a special Object used to contain complex types specific to some component of an application and not necessarily understood by the OPC UA software stack. The Extension Object contains the parameters for OPC UA service messages.
Primitive Type:	Structure of: • *NodeID* – The *NodeID* of the data type Encoding Node that describes an encoding mechanism. • Encoding – An enumeration describing how the body is encoded. • Length – Int32 number of bytes in the body. • Body – Series of bytes comprising the data in the Extension Object.
How It's Used:	The Extension Object is an all-purpose data type that can be used to carry any kind of data including a proprietary protocol or even an audio or video file.
Why it's important:	The Extension Object contains the parameters for a service Message such as the parameters for the Read Service.

Table 10 – Extension Object Derived Type

RequestHeader	
Description:	The Request Header is the header that a Client attaches to every service request message.
Primitive Type:	Structure of: • *authenticationToken* – The token returned by the Server during the Client authentication process. • timeStamp – The time the Client issued the request. • *requestHandle* – The Client's reference ID for this request. See OPC UA Part 4 Specification for other components of this structure.
How It's Used:	The Request Header is used on every service call to pass basic information to the Server.
Why it's important:	The Request Header has the token that indicates that this is a valid request from an authenticated Client.

Table 11 – Request Header Derived Type

ResponseHeader	
Description:	The Response Header is the header that a service attaches to every Service Response message.
Primitive Type:	Structure of: • *timeStamp* – The time the service issued the response. • *requestHandle* – The ID the Client assigned to the original request. • *serviceResult* – A structure of status data indicating if the service was completed successfully, unsuccessfully, with warnings, or in an uncertain state. See OPC UA Part 4 Specification for other components of this structure.
How It's Used:	The Response Header is used on every service call to pass basic information about the status of the request to the Client.
Why it's important:	The Response Header contains very detailed status and diagnostic information about the status of the Client's request.

Table 12 – Response Header Derived Type

ApplicationDescription	
Description:	The Application Description provides descriptive data that describes an OPC UA application instance.
Primitive Type:	Structure of: - *applicationURI* – A unique reference to a particular instance of the device or application in the form "urn:Hostname:vendor:productname". The host name may be the host name of the computer or the IP Address of an embedded device - *productURI* – A reference to the product description: either a URL or a string description of the product like "ACME Inc. 32bit Valve Block". The product URI is unique to the product from that manufacturer - *applicationType* – Enumeration specifying if this device is a Client, Server, or Discovery Server See OPC UA Part 4 Specification for other components of this structure.
How It's Used:	The Application Description is provided to the Server when the Client creates a session. The Server's Application Description is returned to the Client as part of the Endpoint description returned in a **GetEndpoints** request.
Why it's important:	The Application Description provides important information about the Client or a Server application to the other device. It describes what kind of device it is (Client or Server), what the globally unique URI is so that it can be distinguished from other devices, and the product name or the reference to the product.

Table 13 – Application Description Derived Type

endpointDescription	
Description:	A description of an endpoint in an OPC UA Server.
Primitive Type:	Structure of: • *endpointURL* – The URL the Client can use to access this endpoint • *applicationDescription* – The Server's Application Description • *applicationInstanceCertificate* – The certificate issued to the Server by the Certification Authority (CA) • *securityPolicyUri* – The URI describing the security policy used on this endpoint See OPC UA Part 4 Specification for other components of this structure.
How It's Used:	The Endpoint Description is how the Server describes to a Client how to reach an endpoint, what its application is, what security is in place, and what functionality is supported on that endpoint.
Why it's important:	A Client uses the *endpointDescription* to determine if it is able to access the Server on a particular endpoint.

Table 14 – Endpoint Description Derived Type

LocaleID	
Description:	The *LocaleID* specifies the local language using ISO standards.
Primitive Type:	A simple string composed of a language component and a country/region component where the country/region component is always preceded by a hyphen.
How It's Used:	OPC UA includes the ability to display localized text in the language of the locale. The *LocaleID* field is part of the localized text structure which identifies the local language.
Why it's important:	Internationalization is built into OPC UA. When creating a session, a Client can specify a list of *LocaleIDs* that can be used to localize text string during that session. The Server picks the first *LocaleID* in the list that it supports.

Table 15 - LocalID Derived Type

ReadValueID	
Description:	The *ReadValueID* is the structure that describes an attribute to Read in a Read Attribute service Request.
Primitive Type:	Structure of: • *NodeID* – The target NodeID for the read request. • *AttributeID* – The target attribute to read. • *indexRange* – The elements of the attribute to read and return to the Client (Null if the attribute is not an array). See OPC UA Part 4 Specification for other components of this structure.
How It's Used:	An array of *ReadValueID* structures is passed to the OPC UA Slave during the Read Attribute Request message.
Why it's important:	*ReadValueID* is the main structure used to request transfer of attribute and Property values from a Slave to a Client.

Table 16 - ReadValueID Derived Type

Hints for Reading the OPC UA Specification

Part 3 of the OPC UA specification discusses data typing in some detail. There is an early section in Part 3 that describes type definition nodes in general and a later section that describes data typing in particular. The first section gives only a paltry description of data typing while devoting most of its coverage to Object typing and providing an explanation of the difference between Object instances and Object type definitions. The latter sections of Part 3 do discuss data typing but focus mostly on a concept called the data dictionary.

For more information on the *BaseDataType* definition and how it integrates in an address space please see OPC UA specification Part 5, Information Modeling.

DATA ENCODINGS

What's Interesting?

What's interesting about how OPC UA encodes messages is how the encoding is selected. There is more than one way for a Client to select which encoding is used. When a Client initially detects an OPC UA device and requests its endpoint descriptions, the Server returns the endpoint details with the encodings offered on each endpoint. The Client can choose to use the endpoint that best serves the application. A Client that is a front end for a database or some other IT-like application can choose XML (eXtensible Markup Language) encoding if the Server offers it. A Client that is going to be crunching data through some algorithm may choose UA Binary encoding.

But that's not the end of the story. Even though the endpoint the Client selected offers one encoding, a Client can request another encoding. When forming the OPC UA service message requesting the data, a Client can request a specific encoding that is different from the standard encoding on that endpoint. If the Server supports it, the Client can read attributes and properties all day long using one encoding and for other specific variables and properties, get the data in another encoding.

That's an interesting mechanism not found in other technologies.

What Do You Need to Know?

With every networking technology, eventually the bits have to be put on the wire serially: one bit after another. The process of translating every integer, byte, and floating point data value into a series of sequential bits is known as encoding.

There are two encodings available in OPC UA: UA Binary and XML/Text. The binary encoding is provided for smaller-resourced embedded devices while the XML encoding is provided to provide ease of integration with IT applications.

In most, but not all devices, the end user will select the encoding in combination with a transport layer that meets the requirements of a specific application. In most cases, this decision will be made based on the devices or applications being connected. Embedded devices will most likely only support UA Binary, while OPC UA Client applications on more powerful computers will support both XML/Text and UA Binary.

Advantages of OPC UA Over Other Technologies

Most technologies have a single encoding that is fixed and unchanging. OPC UA is one of the few technologies that provides multiple encoding options and expands the breadth of devices that can be connected to include both factory floor and IT devices and applications.

OPC UA is also one of the only technologies that allows the end user to make technical decisions on implementation (like encoding and transport) at deployment. There are tradeoffs that have to be made in every application, and typically those tradeoffs are made during device design. OPC UA is the only technology to allow that tradeoff to be managed at deployment.

What Are the Details?

Encodings in OPC UA define a standard way of representing data on the wire. Before taking data in an application and sending it to another node, it must be represented in a way that the target system can understand it. There are two data formats defined, called mappings by OPC UA: UA Binary and XML. The mappings define how to represent data for public consumption.

The selection of a data encoding is an important application consideration. The data encoding affects the system performance and how easy it might be to pass data to other systems. Data encoded as XML / Text offers reduced performance, but the XML values can be easily passed to many other systems. Data encoded as OPC UA Binary offers higher performance and is best for applications where data is processed to compute some result.

OPC UA data encoding relies on having clearly defined encoding rules for each of its primitive data types. All structures and messages are encoded by serially encoding each of the component primitive data types. All fields must be encoded even if the current value is Null. The OPC UA specification is very clear on how to encode each primitive data type as a value and as a Null value.

XML Encoding

XML is a meta-markup language. That means that data in an XML document is surrounded by text markup that assigns tags to the data values. Each data value together with its distinguishing tag name is an XML element, the basic, defining unit of an XML document. The entire collection of elements forms the XML document.

OPC UA XML encoding is a mechanism for transmitting data values as ASCII values with tag names that identify the transmitted data. The OPC UA XML Schema defines the contents of the XML files and how every primitive data type is encoded.

All OPC UA primitive data types in an XML Schema use a prefix of "xs:" prior to the type. In the following snippet from a schema, the UnitCount field, a property with a data type of 6 (integer), is defined thus:

<xs:element "UnitCount" type="xs:int" minOccurs="0" />

An XML encoding is specified by the OPC UA Profile for the Server device or by the request message from the Client device. This means that a Client can request a particular variable be delivered as binary even though the OPC UA Profile specifies that data is encoded as XML. This is particularly useful for large data transfers.

Binary Encoding

Many low resourced devices may be designed to only support OPC UA Binary Encoding. Binary data encoding is a mechanism for transmitting the least amount of data on the wire. Data is formatted simply as a series of bytes. Int32, for example, is encoded as two bytes in little endian format (least significant byte appears first in the stream).

Binary Encoding is more efficient than XML/Text and able to transfer data with better performance than XML/Text. OPC UA Binary packs data into smaller packets and is generated faster with less code than XML/Text.

The OPC UA specification defines how every primitive data type is transmitted on the wire using binary encoding.

Hints for Reading the OPC UA Specification

Part 6 of the OPC UA specification provides a very good and very detailed description of how each of the primitive data types in OPC UA are mapped with special attention to how Null values are mapped. It contains special sections describing the mapping of arrays, structures, and messages.

SECURITY

Disclaimer

Information security is a very complex and expansive topic, enough to warrant an entire book let alone a small chapter. The scope of this chapter is purposely constrained, to fit within the pages allocated. It focuses solely on the OPC UA Secure Conversation protocol, the mapping expected for most embedded automation devices. Another protocol mapping, for example, is WS Secure Conversation, the security protocol used with IT applications, which behaves similarly to OPC UA Secure Conversation. The reader is free to peruse Part 6 of the OPC UA specification to discover how WS Secure Conversation and other security mappings are implemented.

This chapter is specific to OPC UA Secure Conversation with the UA TCP transport. Serialization is not included, but binary serialization always accompanies the implementation discussed here. OPC UA Secure Conversation security with UA TCP transport and binary encoding is known as *UA TCP Secure Conversation with binary encoding*.

What's Interesting?

To meet the requirements for platform independence and support resource-rich Server devices as well as low resource embedded devices in varied application settings, OPC UA modularizes its communications. Unlike most other information communication systems, OPC UA disconnects its request-response messaging protocol from the serialization, security, and transports. An OPC message, like the Read Service, reads the value of an attribute and can be serialized in multiple encodings (binary or XML). It can also be secured with one or more security protocols (OPC UA Secure Conversation or WS Secure Conversation) and be transported with one or more transports (UA TCP, HTTPS or SOAP/HTTP). In most systems, there is no such division between these functions.

It's also interesting how OPC UA splits the functionality for providing a secure message channel between two applications from the mechanism, initiating and maintaining a "conversation" with a credentialed and authorized user. The Secure Channel services initiate and maintain a long-running communications link that moves messages with confidentiality and integrity. The session services use the Secure Channel services to initiate and maintain a connection with a user that has appropriate credentials and authorized access to specific resources.

Resource authorization for specific entities is not a "go/no-go" decision. Users can be authorized for access to Server resources on a system-wide basis. Other users can be authorized to only perform specific actions on specific Server resources. Resource authorization can be as fine-grained as the system administrator desires.

Unlike many other technologies, OPC UA makes a distinction between Client applications and users. A Server may authorize a connection with a Client application and create a communication channel, while not authorizing a connection with a particular user of that Client application. Applications and users are authenticated and authorized separately.

Finally, auditability. The ability to audit who did what, when, and how is integral to maintaining system security. OPC UA builds that kind of tracking capability and auditability into the base system. The audit trails built into OPC UA can ensure the system is working as intended, detail which users are accessing resources or initiating specific actions, and alert administrators to attempts to compromise the system.

What Do You Need to Know?

To begin a discussion of computer security in general and on OPC UA security in particular, it is important to understand some basic terms:

Public Key – A series of bytes which form a key which the owner makes available to everyone who requests it.

Private Key – A series of bytes which form a key that is kept private by the owner and never released to anyone else.

Digital Certificates – A sequence of data bytes that functions like your driver's license. The Digital Certificate verifies that you are who you say you are. There are many components to a Digital Certificate, including: the name of the algorithm and the organization that created it, the owner's public key and the dates it is valid. X.509 refers to the most popular certificate standard. You will often encounter the term "X509 Certificates." You will also encounter the term "DER Certificates," which refers to the method for encoding certificates as a binary series of bytes.

Certification Authority (CA) – An organization that creates and distributes Digital Certificates. The CA creates the public and private keys that are associated with the certificate owner. The CA often encrypts a portion of the certificate with its private key (i.e. signs it) to assure anyone that the CA did create the certificate. Of course, sending a certificate to a receiver is only effective if the receiver knows the CA is an honest and reputable certificate provider.

Digital Signature – Also called "signing," a digital signature is a small series of bytes that result from processing a larger series of bytes through an algorithm. The resulting smaller series of bytes are encrypted with the owner's private key. Using the owner's public key and validating the result with the sender's same algorithm, the receiver can decode the encrypted result and verify that the true owner "signed" the document. Signing a document or message guarantees the integrity of the portion of the message signed with the owner's private key.

Public Key Encryption – An encryption process in which private and public keys are exchanged to sign and encrypt messages.

PKI (Public Key Infrastructure) –The set of hardware, software and policies needed to manage certificates, keys, access lists, and keys used in Public Key Encryption.

Authentication – Authentication means to verify who the sender is. Usually that means validating the sender's X509 certificate and verifying that the certificate is currently valid, as well as signed by a reputable and trusted Certificate Authority.

Authorization – Authorization is the process of validating access to a resource. Once a sender is authenticated, the sender must be granted access to resources by the receiver. Authorization can be accomplished using a list of trusted names, a user name and password, or any other reliable mechanism.

RSA – RSA is a very popular public key cryptography algorithm. RSA refers to the initials of the three designers of the algorithm: Rivest, Shamir, and Adleman. RSA, with its variously sized key lengths, is used in OPC UA and other popular and secure protocols.

SHA – SHA is a series of public key cryptography algorithms published by the National Institute of Standards and Technology (NIST). SHA algorithms are also used in OPC UA to sign and encrypt messages.

Auditing – The recording of all actions, activities, users, resources and more in a system. In OPC UA, auditing is incorporated as a normative part of the specification, as a mechanism for the system administrators to identify vulnerabilities and diagnose security breaches.

Symmetric Security – In Symmetric security both the receiver and the sender hold the key to decrypt messages. One encrypts with the symmetric key, the other decrypts with it. It is called "symmetric" since both hold the same key.

Asymmetric Security – In Asymmetric security both sender and receiver have a private key that they keep secret, and a public key that they share with anyone. Messages to the key holder are encrypted with the public key and decoded by the key holder with the private key. Message segments that require verification of origin are signed with the private key and validated using the public key.

OPC UA is designed to counter threats most likely to occur in your manufacturing plant, specific to various types of operating equipment with varying amounts of resources and processing power. Threats like message flooding, unauthorized disclosure of message data, message spoofing, message alteration, message replay, and other attacks are countered by the security procedures built into OPC UA.

Some of the ways that OPC UA counters these attacks include:

- Certificate based authentication and authorization. Digital certificates, generated by a legitimate certificate authority (CA), provide application authentication. Note, though, that certificate management and Certificate Authority are not included in OPC UA.
- Authentication and authorization of applications based on credentials as simple as name/password sets, or as advanced as digital certificates. Servers (as configured by system administrators) provide the list of acceptable credentials to Clients during the connection process.
- Confidentiality using standard, well-known, encryption systems and algorithms such as RSA and SHA.
- Integrity, using digital signatures on critical components of message data.

These mechanisms are completely scalable. A Server with minimal resources, using the low functionality Nano Profile, may implement none of these mechanisms. Other Servers with more resources may implement just some or all.

Most OPC UA Servers will provide an administrator with the means to customize security operations of a Server for that particular site's requirements, application and users. In one application, an administrator could, for example, use the Server in two areas, enabling user/password security authorization in one and digital certificates in the other.

What Are the Details?

OPC UA and your Security Strategy

With the introduction of Ethernet to manufacturing environments and the desire for increasing integration with the corporate enterprise, security has become extremely crucial. The truth is that the threats to our manufacturing systems are real, that they are continually present, and a very difficult problem to solve.

It is important to note that OPC UA is NOT the solution for your manufacturing security. OPC UA is only one component of a Cyber Security Management System (CSMS) that addresses your personnel, site security, policies, and procedures appropriate for you processes and systems. Your CSMS must also cover a "defense-in-depth" strategy—that will provide the right firewalls, intrusion detection, application, and user authorizations and controls.

What Is a Security Mapping?

To accommodate the variety of platforms and operating systems where OPC UA may be deployed, the designers separated the processing of OPC UA Service Requests and Responses from the security, transports and encodings. In OPC UA, the process of applying an encoding, a security mechanism and a transport is termed "mapping." Mapping these operations to an OPC UA service request is illustrated in Figure 19.

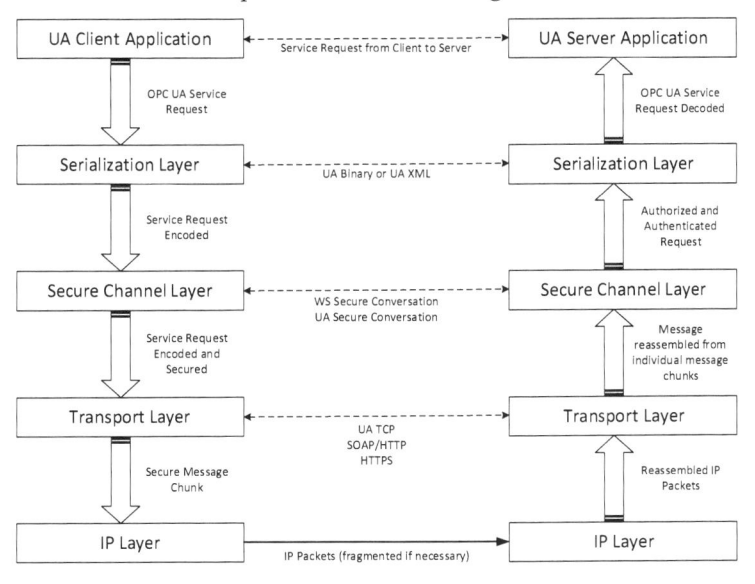

Figure 19 - Service Request mapping Client to Server

In the application layer, the Client generates a service request and Servers process the request. The OPC UA Service/Response protocol is oblivious to any security, encodings or transport messaging. There is no change to the request or response processing no matter what platform or security infrastructure is being used.

In the serialization layer, the service request is encoded such that receiver may efficiently decode and understand the request. Two serialization mappings are available in OPC UA: binary encoding and XML encoding. Binary encoding is used with lower resource devices, which lack the ability to process the volume of characters required for XML encodings. XML encodings are used with IT and enterprise systems to transfer information.

When the serialization is complete, the security mapping is applied to the resulting message string. At the present, OPC UA provides two security mappings: WS Secure Conversation and UA Secure Conversation. UA Secure Conversation mapping is the subject of this chapter.

When serialization is complete and mapping has been applied, the resulting data is ready to be turned into a byte stream and transmitted over IP, the Internet Protocol. The transport layer creates that byte stream through an OPC UA TCP transport with Secure Conversation, then is integrated with a Secure Channel layer. Finally, the TCP Header and Trailer are added to complete the byte stream.

This process of taking a generic OPC UA service request and mapping it to an encoding, a security protocol and a transport, make OPC UA uniquely flexible and adaptable. When new encodings, security protocols and transports are introduced in the future, they can be easily inserted into the same structure.

Security Building Blocks

OPC UA defines a number of entities that are integral to the security infrastructure known as UA TCP transport with OPC UA Secure Conversation. The following tables define these entities:

APPLICATION INSTANCE CERTIFICATE

What It Is:	An Application Instance Certificate (AIC) is a digital certificate created by a CA that represents an application. The AIC is the official document validating the application to the Client.
Structure and Important Components:	• Serial Number – A unique identifier for this certificate • Signature – The signature of the issuing CA validating the certificate • Valid From/To – These fields specify the dates that the certificate becomes valid and when it expires • Subject Name – Typically the Product Name • Application URI – A generic string that provides identification • Public Key – The owner's public key, which can decrypt messages generated by the owner's private key
How It's Used:	When certificates are enabled, the AIC is used for both authorization and authentication of applications and users.
Why It's Important:	The AIC is the main component of the entire Public Key Encryption strategy. If the AIC is generated by a known and reputable CA, then the receiver knows the sender device is "who" it claims to be (and what public key can be used to validate signatures and messages encrypted by that sender).
	The AIC is transferred between a Client and Server during both channel and session configurations.

Table 17 - Digital Application Interface Certificate

SOFTWARE CERTIFICATE

What It Is:	A Software Certificate is a digital certificate that represents an application's software. The AIC is the official document describing the specific functionality of the software.
Structure and Important Components:	(Same components as the AIC, but also includes) • Vendor Name – the product vendor name • Profiles – The OPC UA Profiles supported by the product • Conformance Units – The set of Conformance Units approved by OPC UA certification • Product URI – A generic string that provides product identification • Digital Signature – The owner's signature, validating the certificate
How It's Used:	A Client uses the Software Certificate to determine if this product has the required functionality for the Client's application.
Why It's Important:	The Software Certificate is how a Client knows with certainty that the product meet its functionality, security and operational requirements.

Table 18 – Digital Software Certificate

MESSAGE SECURITY MODE

What It Is:	The Message security mode defines what security should be applied to a particular endpoint.
Structure and Important Components:	The Message security mode is an enumeration that specifies: • No Security – Messages are neither signed nor encrypted • Signed – All messages are signed but not encrypted • Signed and Encrypted – All messages are both signed and encrypted
How It's Used:	The Message security mode is part of the endpoint description returned to the Client on a Get Endpoints message. If the Client chooses to access the Server on that endpoint, it must use the security mode specified in this enumeration.
Why It's Important:	The Message security mode specifies how rigorous security is on a given Server endpoint.

Table 19 - Message Security Mode

MESSAGE SECURITY POLICY

What It Is:	The Message Security Policy defines what security algorithms are applied to messages on a particular endpoint.
Structure and Important Components:	The Message Security Policy is a URI that describes an OPC UA security Profile. The security profile details what security algorithms are applicable under this policy.
How It's Used:	The Message Security Policy is part of the endpoint description returned to the Client on a Get Endpoints message. If the Client chooses to access that Server on that endpoint, it must use the security algorithms (specified in the profile) corresponding to this policy.
Why It's Important:	The Client and Server must use the same security algorithms to verify signatures and decrypt messages. The security policy parameter ensures that both Client and Server use the same algorithms.

Table 20 - Message Security Policy

USER TOKEN POLICY

What It Is:	The User Token Policy describes the credentials that a Client must provide to identify a user to the Server.
Structure and Important Components:	The User Token Policy is a structure with a number of components, including:
	• Policy ID – A String naming the Policy
	• Token Type – An enumeration specifying one of the following:
	• Anonymous – No identification is required. Any user can use the Client Application to access the Server
	• Username – A Client must provide the name and password for the User
	• Certificate – A Client must provide the user's digital certificate for identification
	• Issued token – a WS security token (only used for WS security)
	• Security Policy URI – An information string that specifies how the User token is secured. If not present, the User token is secured by the security policy of the endpoint
How It's Used:	The User Token Policy is part of the endpoint description that is returned to the Client on a Get Endpoints message. The Server returns an array of User Token Policies that it can accept on that endpoint. The Client must choose one of those Token Policies to use on that endpoint.
Why It's Important:	The User Token Policy specifies how rigorous the user credentials must be for a user to access Server resources.

Table 21 - User Token Policy

SESSION ID

What It Is:	The Session ID is a Node ID in the Server's address space that identifies the session. It is the non-public identifier for a unique conversation between a single user and the Server Application.
Structure and Important Components:	The Session ID is a Node ID in the address space and is specified by the same Node ID structure used to specify all other node IDs.
How It's Used:	The Session ID is returned to the Client in response to the Create Session request.
Why It's Important:	The Session ID represents the conversation between the Server and a particular user in the address space of the Server.

Table 22 - Session ID

USER IDENTITY TOKEN

What It Is:	The User Identity Token (UIT) is the credentials a Client passes to the Server to identify the user operating the Client application.
Structure and Important Components:	The UIT is a variable type field that depends on the token type. It is selected by the Client from the list of User Token Policy array (received in the endpoint descriptions). It can be a name and password, a certificate, or a WS-Certificate.
How It's Used:	The Client provides the UIT to the Server when it issues the Activate Session service.
Why It's Important:	The UIT is what the Server will use to identify the user, and determine what level of access it has to the Server Address Space and other resources.

Table 23 - User Identity Token

CHANNEL SECURITY TOKEN

What It Is:	The Channel Security Token (CST) is a token that represents how a channel is secured.
Structure and Important Components:	• Channel ID – The unique identifier of the channel • Security token ID – A byte string representing the keys that are being use to secure the channel. The security token ID is a 2-byte field when Open Secure Conversation is used with UA TCP • Create time • Lifetime – The lifetime of the CST. The CST is invalid when that lifetime expires
How It's Used:	The CST is returned to the Client in response to the Create Channel request. The token ID is included in every UA TCP message to identify the channel.
Why It's Important:	The Channel Security Token identifies to the Server the channel used to process a message.

Table 24 - Channel Security Token

SESSION AUTHENTICATION TOKEN

What It Is:	The Session Authentication Token is a value that associates a session (user conversation) to a particular communication channel (secured communication link).
Structure and Important Components:	The SAT is an "opaque identifier," meaning that it is a series of bytes that have no specific interpretation.
How It's Used:	The SAT is returned to the Client in the response to the Create Session request. The Client sends this value with every service request, which allows the Server to verify that the sender of the service request is the same as the sender that created the original session.
Why It's Important:	The SAT is key to ensuring that a communications session between a Client and Server is not hijacked by a 3rd party.

Table 25 - Session Authentication token

The next sections describe the operation of UA TCP Secure Conversation as a Client connects to a Server and how the entities in the figures above are used.

SECURITY AND THE GET ENDPOINTS SERVICE REQUEST

The first service requested of a Server is the Get Endpoints service. A Client issues a Get Endpoint service request to identify an endpoint in the Server which meets its functionality and security needs.

The Server returns an array of endpoint descriptions, each of which include the Server's Application Instance Certificate, and other security information for each endpoint. Figure 20 illustrates the security information returned in the Get endpoints request.

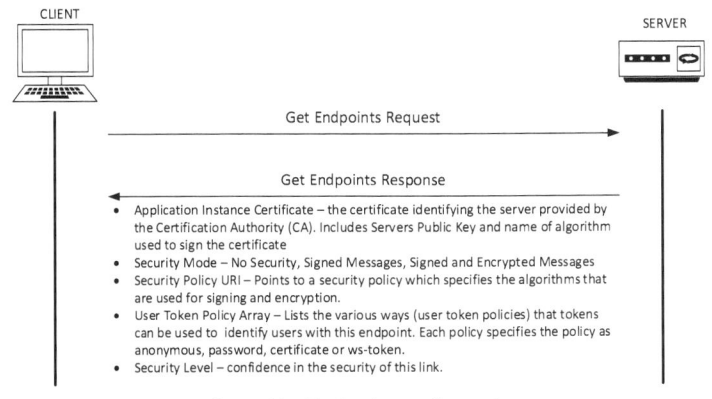

```
CLIENT                                                              SERVER

         Get Endpoints Request
         ─────────────────────────────────────▶

         Get Endpoints Response
         ◀─────────────────────────────────────
         • Application Instance Certificate – the certificate identifying the server provided by
           the Certification Authority (CA). Includes Servers Public Key and name of algorithm
           used to sign the certificate
         • Security Mode – No Security, Signed Messages, Signed and Encrypted Messages
         • Security Policy URI – Points to a security policy which specifies the algorithms that
           are used for signing and encryption.
         • User Token Policy Array – Lists the various ways (user token policies) that tokens
           can be used to identify users with this endpoint. Each policy specifies the policy as
           anonymous, password, certificate or ws-token.
         • Security Level – confidence in the security of this link.
```

Figure 20 - Endpoints and security

SECURITY AND THE UA TCP TRANSPORT PROTOCOL

UA TCP is the transport protocol used with OPC Secure Conversation. UA TCP provides the low-level link for messages moving between a Client and a Server. UA TCP is a simple protocol which allows the Client and Server to exchange basic link exchange information—like buffer sizes, maximum message sizes, and the maximum chunk count. Instead of moving entire messages, UA TCP moves chunks, or message portions of varying sizes. These chunks are each fully secured by UA Secure Conversation (discussed in the next section).

The UA TCP transport protocol has the following responsibilities:

1. Opening a low-level communication link with Clients.

2. Passing message limits to the Client so it is aware of any Sever limitations regarding message sizes and available buffer sizes.

3. Chunking and de-chunking messages as they are sent and received.

The format of UA TCP messages is illustrated in the next section.

SECURITY AND OPC UA SECURE CONVERSATION

Once both the Client and Server understand the link capabilities, OPC UA Secure Conversation protocol (one of OPC UA's security mappings) establishes the Secure Channel. The Secure Channel is a long term logical connection between a Client and Server, ensuring the confidentiality and integrity of messages exchanged. Confidentiality is attained when an endpoint implements encryption, while integrity is attained when the endpoint implements message signing.

OPC UA Secure Conversation is a mapping protocol. It is a "mapping" because it maps security information onto UA message chunks prior to transmission, and validates message chunks as they are received. OPC UA Secure Conversation maps identically onto the UA transport protocol's message structure, as shown in Figure 21.

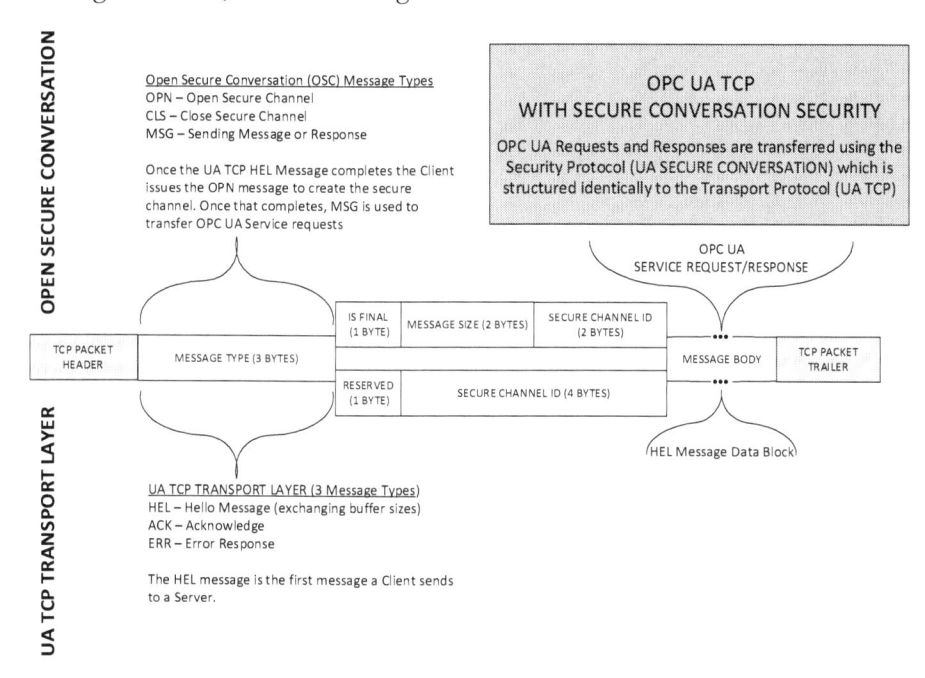

Figure 21 - Secure conversation / TCP transport (identical structure)

The UA Secure Conversation mapping protocol has the following general responsibilities, though they will vary slightly depending on the level of security implemented by the system administrator:

1. Open (and Close) the Secure Channel with the OPC UA Client.
2. Validate that the Client Application is on the Trust List as a known and reliable application.
3. Secure individual chunks as they are transmitted to the Client.
4. Validate the security on each chunk as it is received from the Client.

5. Renew the security tokens for the connection prior to expiration to keep the connection alive.

The message body used by UA Secure Conversation wraps the OPC UA service request in three headers and two trailers as illustrated in Figure 22.

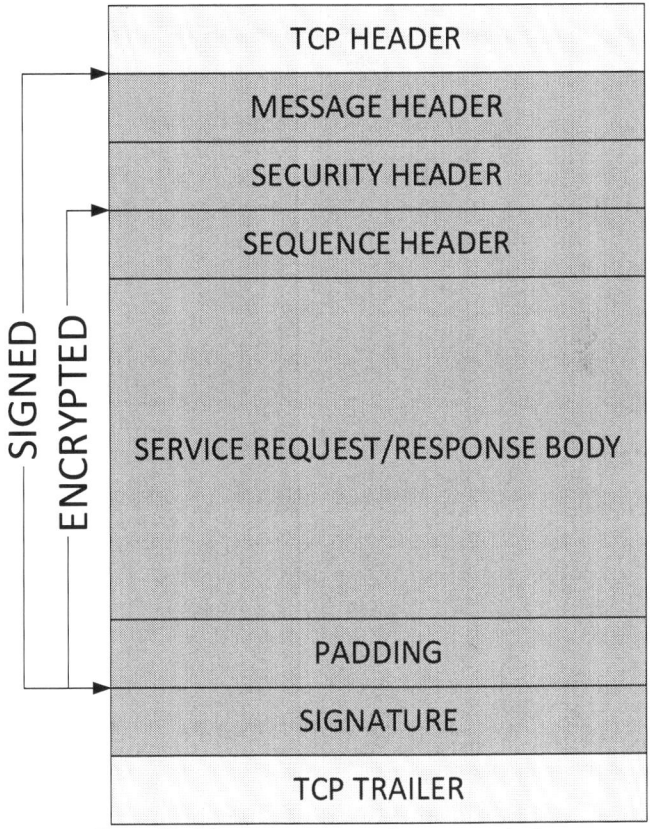

```
                    TCP HEADER
                  MESSAGE HEADER
                  SECURITY HEADER
                  SEQUENCE HEADER

   SIGNED
   ENCRYPTED    SERVICE REQUEST/RESPONSE BODY

                     PADDING
                    SIGNATURE
                    TCP TRAILER
```

Figure 22 - The secure conversation message body

Message Header – The Message Header contains the Message Size, Secure Channel ID, and fragmentation information. It contains information that allows reassembly of the message chunks into complete messages.

Security Header – The Security Header contains a Token ID which references the keys used to create the Secure Channel. The Token ID must be the same Token ID returned to the Client (when the Server created the Secure Channel from the Client's security information).

Sequence Header – The Sequence Header contains a monotonically increasing sequence number which ensures that the header will always be different from chunk to chunk.

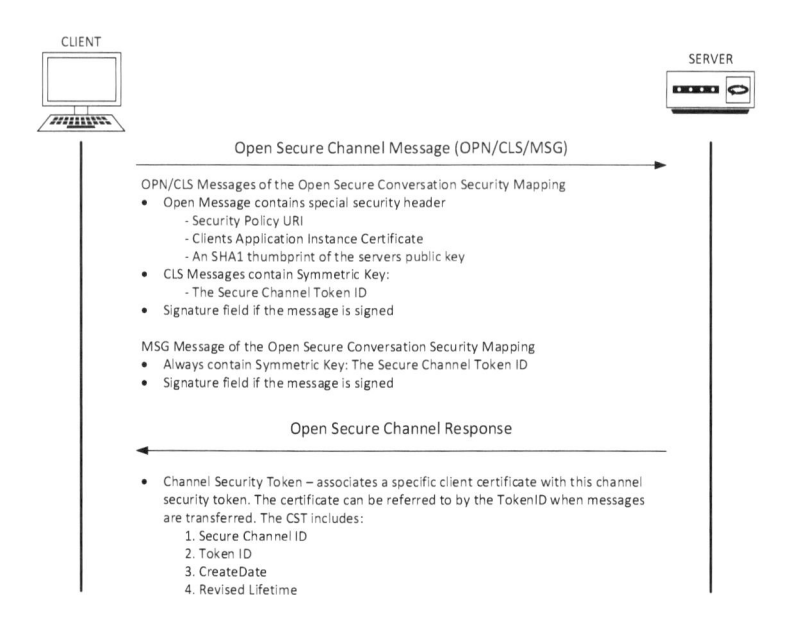

```
CLIENT                                                                    SERVER

         Open Secure Channel Message (OPN/CLS/MSG)
         ──────────────────────────────────────────►

         OPN/CLS Messages of the Open Secure Conversation Security Mapping
         • Open Message contains special security header
              - Security Policy URI
              - Clients Application Instance Certificate
              - An SHA1 thumbprint of the servers public key
         • CLS Messages contain Symmetric Key:
              - The Secure Channel Token ID
         • Signature field if the message is signed

         MSG Message of the Open Secure Conversation Security Mapping
         • Always contain Symmetric Key: The Secure Channel Token ID
         • Signature field if the message is signed

                      Open Secure Channel Response
         ◄──────────────────────────────────────────

         • Channel Security Token – associates a specific client certificate with this channel
           security token. The certificate can be referred to by the TokenID when messages
           are transferred. The CST includes:
              1. Secure Channel ID
              2. Token ID
              3. CreateDate
              4. Revised Lifetime
```

Figure 23 - OPC secure conversation and security

SECURITY AND THE SESSION SERVICES

The UA transport layer and the Channel Services integrated with UA Secure Conversation provide a secure communications link between the Client application layer and the Server application layer. Messages can be securely transferred, with confidence in their confidentiality and integrity over these layers. What's missing is the security of knowing who the user is, if the user is trusted, and that the same Client that created the Secure Channel is also the Client passing the user information to the Server. The job of the session services is to provide trust in the user accessing Server resources.

Creating a secure session with a trusted user is a two-step process. In the first step, a Client requests a session using the Create Session service. Create Session validates the user's digital certificate and compares it to the certificate used to create the Secure Channel. It generates a Session ID and a Secure Authorization Token (SAT) used to identify the session. The Session ID—called a Node ID—is added as a node in the address space, while the SAT is used in the header of all subsequent messages to identify the specific session to which that message applies.

Figure 24 details the security parameters that are transferred in the Create Session request and Create Session response. The Client can validate that it is receiving a response from the requested Server by validating the signature in the Server response. The Server signs the byte string composed of the Client's certificate and the one-time byte string, called a nonce, which the Client provides in the Create service request.

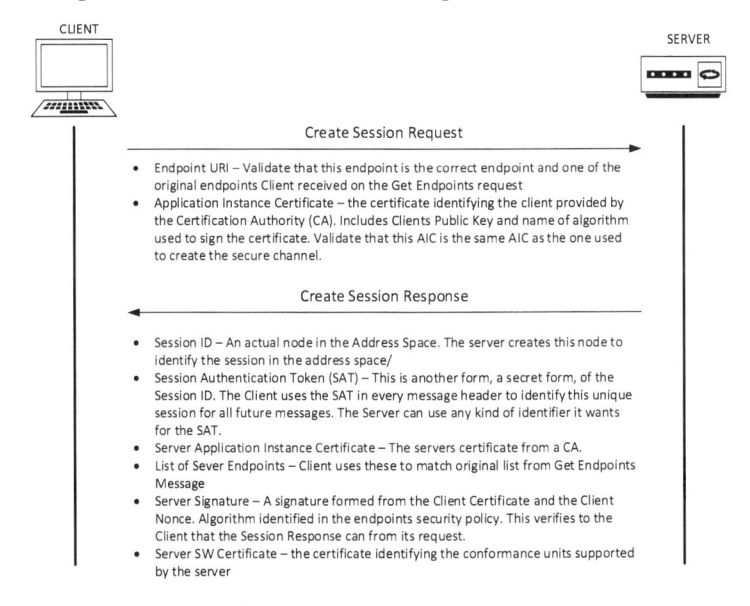

```
CLIENT                                                              SERVER
  [computer]                                                        [server]

              ─────────────── Create Session Request ──────────────▶

              • Endpoint URI – Validate that this endpoint is the correct endpoint and one of the
                original endpoints Client received on the Get Endpoints request
              • Application Instance Certificate – the certificate identifying the client provided by
                the Certification Authority (CA). Includes Clients Public Key and name of algorithm
                used to sign the certificate. Validate that this AIC is the same AIC as the one used
                to create the secure channel.

              ◀────────────── Create Session Response ──────────────

              • Session ID – An actual node in the Address Space. The server creates this node to
                identify the session in the address space/
              • Session Authentication Token (SAT) – This is another form, a secret form, of the
                Session ID. The Client uses the SAT in every message header to identify this unique
                session for all future messages. The Server can use any kind of identifier it wants
                for the SAT.
              • Server Application Instance Certificate – The servers certificate from a CA.
              • List of Sever Endpoints – Client uses these to match original list from Get Endpoints
                Message
              • Server Signature – A signature formed from the Client Certificate and the Client
                Nonce. Algorithm identified in the endpoints security policy. This verifies to the
                Client that the Session Response can from its request.
              • Server SW Certificate – the certificate identifying the conformance units supported
                by the server
```

Figure 24 - Create Session security

In the second step, the Client issues the Activate Session request. The Activate Session is when the user is examined. If the request came from the same Client creating the session and the user credentials are legitimate, the user credentials are passed to the Server's application layer. The Server can validate the user against a trust list or use some other mechanism to determine if the user can be trusted and associated with the secure session. ***The process of determining if the user can be trusted is application specific and beyond the scope of OPC UA.***

A Server performs a number of validations when it receives an Activate Session request:

- It confirms that the same Secure Channel is being used for the Activate session as that which was used for the Create session.
- It validates that the Client is the same Client that created the session by validating the Client Signature. The Client creates the signature from the Server Certificate and the last Server nonce (random byte string) is provided to the Client in the Create Session response.
- It validates that the Session Authentication token is the token provided to the Client in the Create Session response.
- It validates the signature that secures the User Identity Token (if a signature is used).
- It validates the User Identity Token—the authorization for the user to access this Client session. The User Identity Token may be anonymous, meaning that

anyone can access the Server application, it may be a name and password combination, or it could be a certificate. Any credentials provided in the User Identity Token are passed to the user application for final validation of the user.

The operations performed on an Activate Session request are illustrated in Figure 25.

```
CLIENT                                                                    SERVER

                          Activate Session Request
              ──────────────────────────────────────────────▶
    • Client Signature of Server Certificate and the Server nonce
    • Client Software Certificates – List of certificates associated with client.
    • User Identity Token – The credentials of the user associated with the Client. Server
      uses this to see if this user can access the Server resources.
    • Token Signature (if token is a certificate)

                          Activate Session Response
              ◀──────────────────────────────────────────────
    • Server Nonce – Client uses this to prove it owns the session in the next call to
      activate session
```

Figure 25 - Activate Session security operations

SUMMARY – READY TO COMMUNICATE!

It's a comprehensive process. A Client begins by inquiring about what endpoints are supported and then opens a Secure Channel over a transport layer connection. The connection process is completed when the user is authenticated and authorized for access to specific resources in the Server. And now, at the end of this process, the user can send secure and reliable OPC UA service messages to the OPC UA Server.

What messages can be sent? That's detailed in a later chapter on OPC UA services.

Hints for Reading the OPC UA Specification

The specifications for OPC UA security are necessarily distributed between Part 2 (The Security Model), Part 4 (services) and Part 6 (Mappings). Part 2 provides the overall theory behind OPC UA security and describes how OPC UA meets the most common security threats. Part 4 describes the channel and session services. Channel services provide the secure communication link between the Client and Server stacks. Session services provide the secure communication link between a user and a Server application. Part 6 describes how messages are serialized, how the WS Secure Conversation and UA Secure Conversation security protocols are applied, and all the available transport layers.

OPC UA SERVICES

What's Interesting?

OPC UA implements a rich set of services that meet the needs of a variety of different application environments and scenarios. However, compared to other components of OPC UA, the Service Set is pretty ordinary. Even so, there is one interesting aspect of services: it is how result codes are processed.

Since multiple OPC UA service requests can be embedded in a single message, the result codes in a Service Response must report both the result of every individual request, as well as the result for the entire service call.

For each request, the status of individual service requests is provided by an individual status code. The status codes appear in the body of the response message for each service result that is reported to the Client. A standard set of status codes are provided for all services, with a specific set of status codes for each individual service.

The status of the entire service call appears in the standard Response Header, included by a Server in every response. This result code indicates if the requested service(s) were fully or partially successful. It may indicate an error condition on one or more of the operations requested in the service request.

What Do You Need to Know?

There are ten standard services provided by OPC UA Server devices. These services are described in Table 26.

SERVICE	DESCRIPTION
DISCOVERY SERVICE SET	The Discovery Service Set provides the services a Client uses to discover connection endpoints in the Server and evaluate the capability of the endpoints to meet the Client application requirements.
SECURE CHANNEL SERVICE SET	The Secure Channel Services provides services to create a secure communication link between the Client application and the service application. The Secure Channel Services authenticates the Client application and creates a long term communication channel that maintains the confidentiality and integrity of messages passed between Client and Server (See Part 4 of the OPC UA specification for more information on how Secure Channel services operate using OPC UA Secure Conversation).
SESSION SERVICE SET	The Session Service Set creates a communication session between a credentialed user and the Server. The Session Service Set validates the integrity of the user credentials prior to creating the session. User credentials include a user name and password, an X.509 Certificate, a WS-token, or "anonymous." A Server endpoint accepting "anonymous" requires no user credentials. All users are accepted. (See the chapter on OPC UA security for more information on how session services operate using OPC UA Secure Conversation).
NODE MANAGEMENT SERVICE SET	The Node Management services provide a mechanism for a Client to create or delete nodes in the Server's address space.
VIEW SERVICE SET	The View Service Set allows the Client to partition the Server Address Space into subsets, which can be acted on as "mini" address spaces.
QUERY SERVICE SET	The Query service set is used to provide bulk data access to the Server.
ATTRIBUTE SERVICE SET	The Attribute Service Set is used to read or write specific attributes of nodes in the address space.
METHOD SERVICE SET	The Method Service Set provides a mechanism to invoke "methods," what OPC UA calls a small logic program that a Server makes available. Parameters, defined as properties of the method, can be passed to the method as program arguments.
MONITORED ITEM SERVICE SET	The Monitored Items Service Set provides the facility for Clients to create, modify, and delete Monitored Items. Monitored Items are entities that generate notifications when attributes or variables change in value or on the occurrence of alarms or events.
SUBSCRIPTION SERVICE SET	The Subscription Service Set manages subscriptions. Subscriptions inform the Server how often to send notification results when a Monitored Item generates a notification.

Table 26 - OPC UA Service Sets

What Are the Details?

This section provides a little more detail on some of the more important OPC UA services that are not discussed in other chapters of this book.

Secure Channel Service Set

The Secure Channel service (SCS) set defined in Part 4 of the OPC UA specification is not always used exactly as Part 4 defines it. The SCS set is defined exactly like each of the other service sets, but in actuality it functions as part of the security mapping for the endpoint. When WS Secure Conversation mapping is selected, the SCS set becomes part of the Web Services messages. When the OPC UA Secure Conversation mapping is selected, the SCS set becomes part of the UA Secure Conversation messaging. It is not used as a standalone service, as you might think, from reading Part 4.

The two most important services in the SCS set are the Open Secure Channel and the Close Secure Channel. These services provide the creation and removal of the secure communication link between the Client and service applications. The Open Channel service validates the credentials of the Client application, and establishes a confidential communication link between the two applications.

The communication link is as secure as the Client endpoint requires it to be. Servers can expose endpoints that require no security, or both encryption and signing. Even if no security is required, the SCS set establishes a communication channel between the Client and Server application.

Note that the SCS only establishes the communication link. The link does not mean that any user for the Client application has access to all the Server's data or resources. The level of access available to a particular user is controlled by the Session Service Set.

Session Service Set

The Session Service Set manages the communications session between one user and the Server application. The session is established in two steps: Create and Activate.

In the Create stage, the session is created. The Server validates that the same Certificate used to create the channel (if the security mode is not "none") is being used to create the session and that the session timeout is appropriate. Two entities are created by the Create Session request. They are the Session ID and the Session Authentication token. The Session ID is a Node ID for the node created by the Create Session service in the Server Address Space. A Client can browse the address space to determine what sessions are currently in progress. The Session Authentication token is a value that associates a session (user conversation) with a particular communication channel (secured communication link). The SAT is an "opaque identifier," meaning that it is a series of bytes that have no specific interpretation. Once it receives the Create Session response message, the Client must provide the Session Authentication token in the message header of every subsequence message.

Node Management Service Set

The Node Management Service Set gives the Client the capability to modify the Server's address space. The Client can create nodes and remove nodes to build an address space that meets its needs. These types of services are present in other industrial protocols, but have seldom been implemented by device vendors. It's an open question as to whether or not they will be needed or implemented in OPC UA Server devices.

View Service Set

Clients use the View Service Set to navigate through the Server's address space. Clients can use the View services to navigate through the Server's entire address space or smaller portions called "Views". A Server makes available View nodes in the address space to identify the views that are supported. A Client can find the Views available by navigating from the main Root Object to the Views Folder, and then to each of the View nodes. A View node is the root node for a view. A View node contains a hierarchical reference to one node of a View, as shown in Figure 26. By definition, all nodes of the View have to be accessible from that starting node.

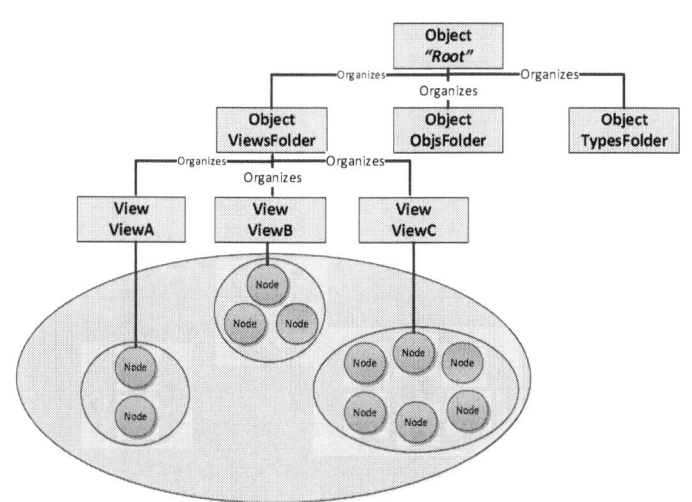

Figure 26 – Root folder and its Object folders

Views are used to limit the scope of operations on a large address space.

The Query service can be limited to only returning data within a View. Subscription services can be limited to notifications of events and value changes within a View.

Browse is the primary service of the View Service Set. Browse allows a Client to discover the references that link each node of a View to one or more other nodes. By discovering these references, a Client can create an internal image of the hierarchical structure of the Server's address space, and navigate to any node in it. Browse can move through the entire address space (the default view) or within a View subset of the address space. The Server limits the references it returns to the Browse service to the requested View. If a node only has references to nodes outside the View, then the Browse service returns a response containing no references.

Attribute Service Set

Every node of an OPC UA Node Class is described by one or more (up to 21) predefined node attributes. Attributes used by all Node Classes include items like the Node Class, the node description, the Display Name, the sampling interval, and its value. The Attribute Service Set is the simplest of all the OPC UA services. It provides the mechanism for Clients to read or write these attributes. A service message contains more than one read or write attribute request.

Access to an attribute can be blocked by the Server. An overall access level defines what level of access all Servers may have to the attribute. For example, write access is denied to read-only attributes. Access may also be denied based on the user authorization level. A user may not have the authorization to read or write a particular attribute. Implementation of the mechanism to validate user authorizations for a particular attribute are outside the scope of OPC UA.

Query Service Set

Whereas the Attribute Service Set provides access to individual attributes, the Query service provides large data services. Query services are designed to return bulk information from the address space in a way that doesn't require the Client device to know anything about the logical schema for the address space. Query is designed for occasional tasks, like populating recipe data in a Server, or returning a large amount of data in a report. Queries operate on the instances of a type definition and have a very sophisticated mechanism for filtering data returned from the query.

Method Service Set

The Method Service Set provides the mechanism to initiate a Method, a small logic program in a Server. Call is the only service provided in the Method Service Set. The methods available in a Server can be discovered by the Browse service, by examining a node's *HasMethod* references.

Arguments can be passed on the execution of a method. The call request service includes input arguments. These arguments must match the preexisting arguments defined for the method in the address space.

Execution of methods is unpredictable, and can vary from an immediate to an indefinite period of time, depending on system status and the logic being executed. Completion of methods can be monitored as a monitored item and notifications can be provided using the subscription service.

Hints for Reading the OPC UA Specification

Part 4, OPC UA services, is a very readable, well-done section of the OPC UA specification. A couple of notes on reading Part 4:

1. Pay attention to the indenting in the parameter lists for the request and responses for each service. Structures are used in these parameter lists and defined elsewhere. Other times, the structures are defined in line.

    For example, in Table 27, "*ItemListX*" is an entity that is defined somewhere else in Part 4, or even in another part of the specification. "*ItemListY*" is defined in line with two components, component A and component B.

ItemListX	…description of Item list x…
ItemListY	…description of item list y…
ComponentA	…description of component A…
ComponentB	…description of component A…

    Table 27 - Example in line data definition

2. There is a standard Request Header for every service request and a standard Response Header for every Service Response. These headers are defined later in Part 4.

3. Some fields used in service requests and service responses are defined in Part 4, while other, very common items are defined in Part 3.

OPC UA IN PRACTICE

It was said that when the great "Shoeless" Joe Jackson was in left field, "Triples went there to die." The same phrase could be used for factory floor data: "the factory floor is where data goes to die." In a lot of ways, it's true. The plant floor is one big morass of physical layers, communication protocols and application requirements that get in the way of moving data easily.

OPC UA isn't going to solve all of that. It really won't ever be solved completely, because there will always be reasons to invent a new technology or deploy something in a manner differently from the way it's deployed in other applications. It's the nature of the physical world. On the factory floor, everything can be neat as a pin in a digital system, but when you have gears, motors, and the constraints of physics, it tends to be less clean and straightforward.

OPC UA can help us interface to that world more easily and simply and make our jobs easier. In this chapter we'll look at some different occupations and speculate on how OPC UA will change the lives of these professionals.

OPC UA FOR CONTROL ENGINEERS

Running a set of machines in a manufacturing plant today is clearly complex. Massive amounts of technology must be coped with and understood. Every day, when you walk into the plant you might be confronted with mechanical issues, electrical issues, environmental issues and, now, IT issues. Essentially, everything that happens around your machines is going to end up in your lap at some point. It's a challenging job!

What's worse now, you're probably responsible for more plant operations than ever before. Everybody runs lean today, asking their people to do more and take on more responsibility. There's always something being upgraded, something fixed, and problems to solve.

So, how does OPC UA help a guy like you? There are several ways.

One is information. You're probably the focal point for many people who need information off the factory floor. Most people haven't the faintest clue about the architecture of the system, what a PLC is, let alone its capabilities. They don't know what's available to them or how to get it. So they come to you, because you're the guy that keeps the machines running and the product moving out the back door.

On the factory floor we've always had a lot of information—but it's been locked up tight. Valves had valuable wear information. Drives had energy data. There are all sorts of non-critical and even some times critical data that are not part of a DCS or PLC. They would be useful, but have previously been hard to find, let alone get.

With OPC UA, all of that changes for you in several important ways. First, the data is easier than ever to find. You can now use an OPC UA Client tool to explore your devices and not only see what kind of data is out there, but also what kind of data format it uses, and other data characteristics–what the computer scientists call meta-data. That's data like the last time the set point was changed, or the units on temperature.

Of course you understand the security risks to your plant. OPC UA is going to help you with that, too (as you learned in the chapter on Security). You'll be able make your information more secure than ever. And with transports like HTTP and HTTPS, you'll be able to more easily and seamlessly move your data to production Servers, maintenance applications, and the cloud.

OPC UA FOR SYSTEM INTEGRATORS

If you're in the System Integration business, doing internal projects for your company or projects for customers all over North America or the globe, you're going to love OPC UA.

System Integration is a challenge—in fact, it's always been a challenge. In the old days the challenge was wiring. You had to buy all that wire, hire big guys to pull it all over a factory, and terminate it, literally making thousands and thousands of terminations. And then you had to check out all those wires, one by one by one, for days on end.

Today it's not the same. Instead of discrete wires, we have networks. You're checking out communications between devices on EtherNet/IP or ProfiNet IO, Modbus TCP, DeviceNet, Profibus DP, Modbus RTU, or some other networking protocol. Not as physically demanding as pulling thousands of wires through wire runs, but pretty challenging all the same, in that you have to understand many more technologies. There's a lot more going on troubleshooting two devices that don't connect versus trying to fix the termination of wire #1721.

But, the biggest challenge by far for the system integrator folks is HMI integration. It's monstrously expensive. Before you can even build the HMI screens, you have to make sure you have all the icons for the process, and, most of all, that you've loaded all the tags.

If there are 15 drives, you have to load all the tag names for each drive and correctly point them to the network device. You'll also need to load all the tag names for the pumps and the photo eyes and the rest of what's needed for the application. Then you have to validate every IP address, DeviceNet Mac Address, ProfiNet IO IP address, or Modbus Register address that you're using. Just like pulling wires, it's a lot of labor, extremely time consuming and costly.

With OPC UA, you're going to expedite this process. A lot.

With OPC UA, you are going to have standard data models for the most common devices on the factory floor (if not all of them). These models will be available to you as XML schemas. You'll be able to load them right into your HMI and your HMI will know all about the devices in your process. Instead of loading the tags for the 15 drives in this system, you'll just provide the IP address of the drives, and it will automatically match up all the data in the drives with the tags in the model.

No more checking out vast numbers of data tags. No more hours of fixing incorrect spellings of tag names. No more simplifying the data you collect to expedite delivery of the HMI.

Vastly simpler. Vastly more productive. OPC UA truly has a lot to offer you if you're in the System Integration business.

OPC UA FOR MANAGERS

One mistake that a lot of managers make is thinking that OPC UA is just another factory floor technology. If they've been around a while, they started with some Modbus, and later, probably did some Profibus DP, or DeviceNet. Then came Ethernet, EtherNet/IP, and ProfiNet IO. It's natural to assume that OPC UA is the next version of factory floor networking. But that would be a mistake.

OPC UA is an information technology, not an I/O protocol. All of those technologies just listed are really good at moving bits from here to there in an automation system. The bits might be valve states, motor speeds or temperatures, but they are mostly inputs and outputs of one sort or another. That's not what OPC UA does. OPC UA is an information technology—it organizes information and provides services that move it where you need it.

As a manager, the amount of information you currently have is not nearly what you'd ideally have. You want to embark on cost reduction. You want to improve raw material quality. You want to optimize your machines.

You want to make your building more efficient and green. You want to launch products more quickly. There are a lot of "wants" and every single one of them necessitates acquiring more data, delivering it to where it can be analyzed, then using the information generated by analysis to deliver better outcomes.

If I were you, I'd be hesitant to bring another "newest and greatest" technology into the facility. You know the costs of new technology. There are learning curves, training issues, and startup problems that always delay your schedule and eat into your budget. It's never as quick and easy as the technologist says it is.

You've probably already implemented a lot of "communication" – those technologies mentioned earlier. They have their place—an important place in automation—but today you really do need to make the investment in OPC UA.

OPC UA is the best way for your team to acquire data and deliver it to systems and people that can make intelligent decisions. When you think about this, there are two advantages that you should consider. 1) Security – this is something that too many managers ignore, as they face day-to-day pressures to deliver and produce. OPC UA is built to implement the level of security you need in your application.

You also are in control of how much security you get. In some applications, you need very tight security, while in others you'll need none. What's different about OPC UA is that you get to decide. Even better, you can change at your discretion what type of security you use, how it will secure the communications link, and how users are authorized.

The other big advantage of OPC UA is that it integrates with the IT side of your business. With support for HTTP, SOAP, and HTTPS, OPC UA is the first technology that really works well with both the factory floor and the enterprise. Every trade group promoting a technology will claim to be the "one and only" technology for you, but OPC UA is the one that comes closest to meeting that claim.

OPC UA FOR IT PROFESSIONALS

If you're an IT Professional, there is a lot to gain from OPC UA. In the past, data on the factory floor was locked up really tight. You had to jump through incredible hoops to pull plant floor data into databases, dashboards, and enterprise applications. And the systems that were built became a house of cards—ready to fall on the next modification to the programmer controller data table, or a Microsoft Windows update.

Pulling data from the factory floor is going to get easier–not great–just easier. There will still be data locked up in lower level systems, buried behind media and protocol barricades that won't be readily available…but, in actuality, that's OK. You don't want access to thousands of plant floor devices; you just need access to the few that can provide you with a gateway to the rest of the data.

A few PLCs are beginning to come equipped with OPC UA. That in itself will be tremendously valuable to you and your internal or external customers. PLCs have a lot of the data that your ERP (Enterprise Resource Planning) systems need, and those systems will be able to access it much easier than previously. We will have to see what transport layers PLCs of the future provide to you, but you can expect at some point for them to support standard IT transports like HTTPS and HTTP/SOAP with the security capabilities that you expect.

Many other devices with valuable data will also be available to you. Motor drives and motor controllers will probably offer OPC UA for access to energy data. Any embedded system with quality data will certainly provide that capability, as will many of the tools that provide diagnostic capabilities.

The world of IT is going to get better in the future. Instead of groaning or throwing your hands up in the air when you have a requirement that needs factory floor data, you'll actually be able to get what you need quickly and easily.

SUMMARY

Now, there are somethings you won't get from OPC UA. I am a cheerleader for OPC UA because I sincerely believe that this technology is revolutionary. As I am not paid to be a cheerleader I can honestly tell you what the real story is and where the bodies are buried.

OPC UA is not going to vastly enhance the configuration process. You're still going to face the usual quirks of how different vendors configure their devices. What will never change is this: the more features a device has, the more complex it will be and the more difficult it will be to configure. A device that has 465 features is going to be light years harder (and require light years more time) to configure than a device that solves a single job and has three configuration parameters.

OPC UA is probably not going to make it that much easier to get data from Programmable Controllers, like Siemens, Modicon, Rockwell, or anyone else's controller. I am open to possibilities, but at the time of this writing there's not much in the way of OPC UA controller offerings. I continue to be impressed by how unwilling controller makers are to keep access to their internals proprietary. Everything will hinge on the object models that are used to provide access to their data. Some will undoubtedly make things increasingly accessible while others will make it more difficult. We will have to wait and see how OPC UA controller access develops, but I am a skeptic as to what extent and how long this development will take.

Lastly, it's going to take a long, long time for OPC UA to substantially penetrate the factory floor, or in the building. It takes years for companies to decide on implementing OPC UA, then actually go through the development process. Once developed, it has to be field tested, certified, and pronounced ready. Next, the sales and distribution channels have to be trained up. Then they have to spread the word to customers. It's a long cycle. But OPC UA will get there.

And it'll be worth the wait.

THREE USE CASES

USE CASE 1: OPC UA MOTOR CONTROLLER

One of the keys to manufacturing in the future is control of energy usage. No one seriously thinks that energy won't cost more in the future. Management of energy usage across an enterprise is a key to practical management of the 21st Century manufacturing enterprise. Energy management is key to shifting resources and production to areas with the lowest cost and sometimes doing that on an hourly if not more frequent basis.

To manage energy usage, you have to know your energy usage. You can't manage energy until you can get a handle on how much of it you're actually using. In today's manufacturing enterprise that means motors. Motors in the US account for a massive percentage of all the energy usage in manufacturing (sometimes up to 70%).

Once you know how much energy your motors are using, then and only then, can you take action to control those costs. If the newer motors in your Alabama plant are more energy efficient than the ones in the Michigan plant you might schedule more production at the Alabama plant. You might only work the night shift in the Michigan plant to take advantage of lower nighttime energy rates. Or you might decide to invest in more efficient motors. None of this is possible unless you have hard data on the energy usage of your plant correlated with your utility rates.

Unfortunately, getting this kind of data from today's manufacturing systems is difficult given the technology that's been available.

Why Use OPC UA?

A simple and effective way to do this is with OPC UA. An OPC UA-enabled motor controller can send energy data from your motor controllers directly to an energy management enterprise application. That's done completely separate from the I/O channel to the machine controller even though it uses the same Ethernet connection and the same Ethernet network.

In Figure 27 the motor controller is connected to both the PLC for machine control and an enterprise Server running an energy application over the same physical Ethernet connection.

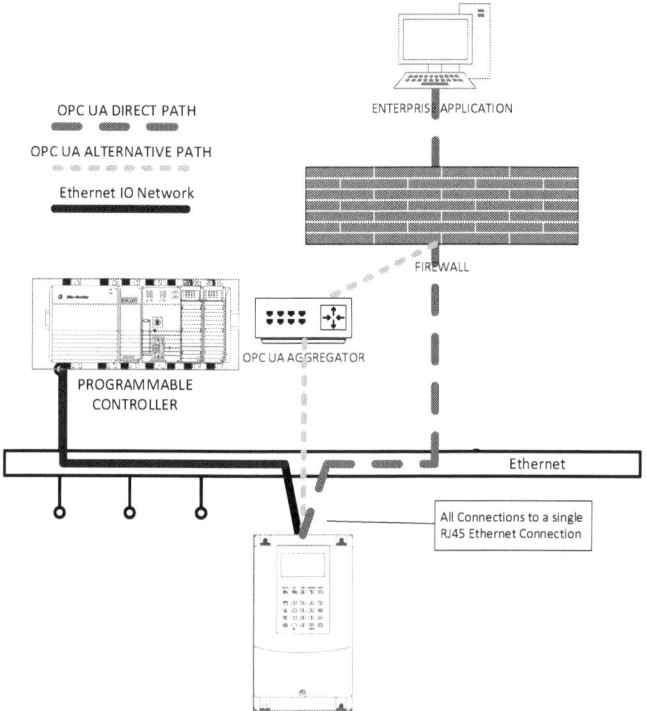

Figure 27 - Motor Controller connected to enterprise

The advantage to using OPC UA in this system are many:

1. The enterprise device can be authenticated meaning the motor controller knows that the device is who it says it is.
2. The application can be authorized meaning the motor controller can know that this application is allowed to access its energy data.
3. The data is encrypted and protected from outside inspection.
4. The data is time stamped. Energy usage can be easily correlated to process changes.
5. The device can be automatically interrogated to learn the data type, limits, name, and units of the data items.

6. Communication problems can be solved much easier due to the simpler configuration.

7. Devices can be more easily replaced

What Are the Alternatives?

The traditional mechanism (Figure 28) for achieving this connection is to move the energy data from the motor controller to the Programmable Controller over the I/O network and then move that data from the Programmable Controller to a PC within the manufacturing firewall. And then, finally from that PC to the enterprise system.

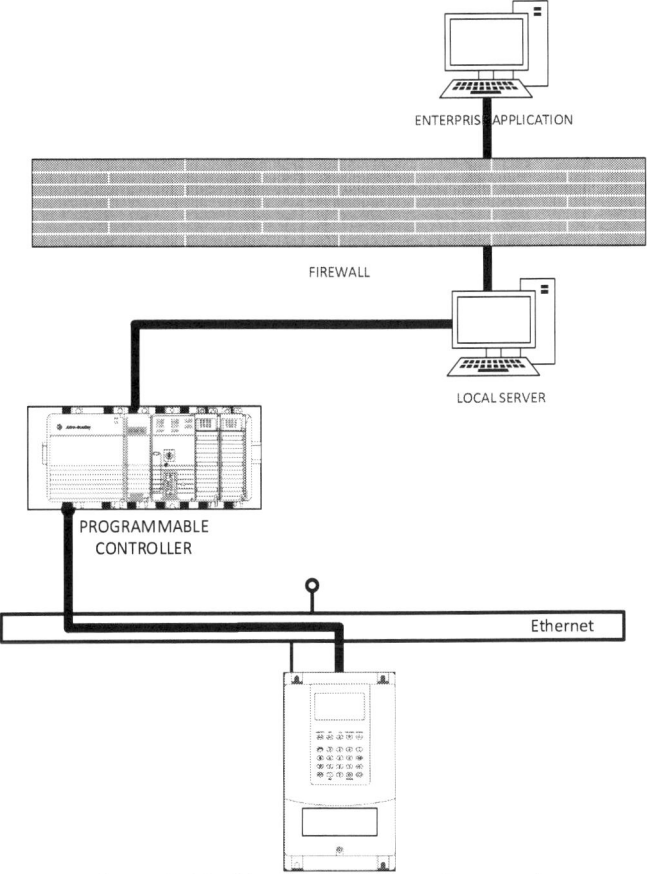

Figure 28 - Traditional factory data to the enterprise

There are significant disadvantages to this approach:

1. There is no meta-data attached to that data. Networked I/O protocols simply move raw data from a device to another device. Any meta-data

like data types, limits, units is lost. When you have devices from different manufacturers the data types, encodings and data formats vary. Data received at the enterprise system is nothing more than a pile of raw data bytes.

2. The Programmable Controller may not support the same data type or precision that exists in the motor controller. The data might get changed as it is moved.

3. Data Integrity is lost. When was that data captured? What is the latency of this system? There is no way to know. It is extraordinarily difficult to coordinate an energy-related event with other real world events in a system with so many moving pieces.

4. Troubleshooting is a nightmare. The complexity is mind boggling. A data item moves from its source through two other systems, each with different data storage mechanisms, sometimes different data types using different protocols.

5. Securing the data is literally impossible. Authorizing and Authenticating users in this system is extremely difficult. Any critical data cannot be protected.

This type of system is a house of cards, a brittle system.

USE CASE 2: OPC UA ELEVATOR MONITOR

Patience is not a virtue that many of us contain. Almost nowhere is that more apparent than when we push the little button and start what always seems to be an endless wait for the elevator doors to slide open. That wait is very important to building operators but it is of vital concern for facilities like hotels where slow running elevator systems can ruin an otherwise excellent guest experience.

Ignore elevator maintenance and slow elevators is what you get. But worse than that, the building owner may unknowingly also be experiencing high energy usage, excessive heating in the drive system enclosure, and excessive noise generation which all lead to a shortening of system life and in excessive cases, safety issues. Knowing that status of elevator operation is critical to providing excellent service to the building owner.

Remote monitoring is critical to providing excellent elevator service and avoiding long term outages that affect building owners and tenants. There are many items to monitor, few of them that are regularly monitored in the one million thirty thousand elevators and escalators in North America. There are hundreds of items to monitor but some of the crucial ones include:

- Travel time
- Energy usage
- Control System Temperature
- System Power Factor
- Door Cycle Times
- Total Operating Hours

Why Use OPC UA?

OPC UA is an excellent solution for providing this type of monitoring for all the same reasons that make UA such an effective tool on the factory floor.

Remote Monitoring – OPC UA is especially effective in situations where a remote piece of equipment must report to a supervisory system when an operational discrepancy is reported. Since UA uses standard Ethernet communications any physical layer including wired, wireless, cellular, and satellite can be used to move OPC UA messages.

Event Handling – OPC UA contains a sophisticated event management system. OPC UA Client devices can select the variables to monitor, select under what conditions the variables values are published, and set the rate at which the remote device's variables are monitored.

Security – For competitive reasons elevator vendors prefer to keep data like operating cycles out of competitor's hands. This is more important if the elevator builder is remotely sending commands to the system to change operating parameters as the system ages, or in response to a high temperature environment, or a utility power charge at a particular time of the day. UA provides excellent security in this kind of situation.

Information Modeling – Using a standard, open Information Model makes it much easier for the elevator manufacturer, their customers, and third-party after-market vendors to build applications that use elevator data. Using standard objects with informative meta-data means that dashboards, hand held applications, and other monitoring systems can be quickly and easily constructed. Manufacturers and customers tracking the frequency of problems can lead to reduced call frequencies, lower maintenance costs, and the avoidance of injuries – all increasing bottom line profits for both the operator and the manufacturer.

Standard Interfaces – Using the standard interfaces (Web Services, HTTP, HTTPS) provided by UA means that the elevator operator can easily integrate the elevator operating data into other systems like the facility's Building Management System providing more complete control over the entire building operation.

USE CASE 3: Building Automation Energy Monitoring

Monitoring energy, water consumption and gas usage is very important to building owners. These items comprise some of the major operating expenses of a commercial building. Unlike a lot of other expenses, they are hard to predict as a factor totally out of the building owner's control, the weather, and directly affects what those costs are. Worse, few building owners have the monitoring systems in place to even attempt to mitigate these costs.

Commercial buildings have various kinds of Building Management Systems. A lot of small commercial buildings like offices, schools, churches, and residences usually have no Building Management System. Setting the wall thermostats is often times the only control that many have. Larger commercial buildings may use a simple control system that sometimes has the smarts to modify settings for nights, weekends, and holidays. These systems may provide some trend analysis but often have no mechanism or a difficult-to-use mechanism to transfer data for more detailed analysis.

There are some commercial buildings that have a very smart, fully automated, very sophisticated Building Management System that can shed generator loads, control backup generators, regulate furnaces and ventilators, and adjust chilled water production. Some of these systems monitor activity levels and regulate systems based on to-the-minute pricing data and weather.

In states like California where water shortages exist, monitoring water usage is now a priority. This is especially true at data centers that host Internet Servers. As California leads the nation in the number of "Server farms," these building owners are under extreme pressure to reduce the extraordinary water usage at these buildings.

No matter what kind of building control system is used, it is not generally designed specifically to monitor, track, and analyze the usage of these types of consumable. And even ones that have that capability aren't the best and lack the ability to combine data from multiple sites to present information as a correlated whole. Few, if any, have the ability to monitor external rates for these consumables and make adjustments in building operation to reduce consumption at critical (expensive) times.

There are sophisticated analysis tools that are specifically designed to do this work but how do we get the data to them?

Why Use OPC UA?

OPC UA is advantageous in this application domain for any number of reasons.

1. Security – It is likely that an enterprise analysis tool will be using utility data to regulate building operation. Commands flowing back to the building from the tool must be secured and OPC UA provides the best security of any alternative.

2. Integration with BACnet – As of this writing the BACnet organization is developing plans to use OPC UA to move BACnet objects. That means that Building Automation vendors can utilize the same BACnet Object definitions for air handlers, chillers, ventilators, and all the other Building Automation devices they use today.

3. Optional Configurations – It is likely that many of today's Building Management Systems (BMS) will implement OPC UA. That means that devices that support mechanisms for transfer to the BMS can use the BMS as a channel to the enterprise.

4. Remote Access – OPC UA provides the best way for remote building data to be integrated with an enterprise analytics tool or even a Building Management System.

What Are the Alternatives?

The alternatives mentioned earlier in this chapter can all be used with Building Management Systems. All lack features found in OPC UA.

Vendors of Building Automation systems may argue that the ideal system configuration is one in which this data is captured by the Building Automation system then moved to these analysis tools. Some of these systems may even support OPC UA in this configuration. That's an interesting alternative but in the end it's going to be found to be more complex and more difficult to integrate.

Direct connection with the device capturing the data is always the ideal way to integrate building data with enterprise analysis tools. That approach provides the fastest integration, the most flexibility, and the lowest risk. By interacting directly with the meters, OPC UA ensures that there is no data loss, no data typing inconsistencies, and the highest level of integration.

A good general principal to avoid the kinds of "brittle systems" that have plagued building owners in the past is to limit the number of systems that your data has to pass through on the way to its destination. The least amount of handling that's done, the better the latency, the more accurate the result, and the lower the risk of System Integration issues.

What's Interesting?

What's interesting about this use case is the amount of building management data that there is and how it is viewed differently in different parts of the country. In California, water and water monitoring is really a top priority. On the east coast it's power. In the South it's responding to ambient temperatures. But no matter what the geographic location or the issue to manage, the problem is the same, moving data to a place where it can be used and turned into information that can be acted on.

COMPETITIVE TECHNOLOGIES

XML

The eXtensible Markup Language, known as XML, is becoming increasingly important in the world of Industrial Automation. It's a perfect storm of functionality and requirements. XML is a data language which communicates by sending files of ASCII characters from one system to another.

XML is verbose – there's no getting around it. It's also simple and human readable. Like it or not, XML is the standard IT folks use worldwide. All of the standard offerings from Microsoft, like Word, Excel and the rest, are XML-based. And it's those same IT standards that are being pushed down to the factory floor.

What Is XML?

XML is a meta-markup language. That means that data in an XML document is surrounded by text markup that assigns tags to the data values. Each data value, when taken together with its distinguishing tag name, is an XML element—the basic defining unit of an XML document. An entire collection of elements forms the XML document.

An element is formed by a start-tag, an ASCII string, and an end-tag. All tags are enclosed in angle brackets like this: <…tag…>. End-tags signify that they are end-tags by preceding the tag name with a slash. A few well-formed XML elements are, for example:

<name> Emily Ward <\name>
< sentence> Where is the family dog? <\sentence>
< temperature> 22.53 <\temperature>

While the names for XML elements have fewer restrictions, XML documents follow a very specific and strict grammar. The grammar specifies where XML elements can be placed, how child elements are specified, how child elements are associated with parent elements, and how attributes are attached to elements.

The markups (elements) allowed in a particular application are defined in an XML Schema. A Schema defines all of the valid elements in a document and allows a generic parser to determine if an XML document is well-formed for a particular application. A document can be well-formed for one application (chemist composition) and invalid for another application (court case). Non-well-formed documents are ignored by a receiver.

How Is XML Used?

XML documents are standard text documents that can be created and edited in any text editor, or in a word processing program like MS Word. Once an XML document exists, it can be transported in any number of ways from a sender to a receiver. In many cases, a device is triggered to transmit an XML document by simply referencing a URL for the XML document.

XML is a good choice for non-control data, especially when that data needs to be sent to different types of applications across various platforms. XML is essentially the lowest common denominator for transferring data. Anything that can decode an ASCII character can parse an XML file. XML is the most prevalent and highly integrated technology for moving data between two non-heterogeneous systems. There is no other technology that is supported across so many platforms and applications. XML parsers are built into browsers, databases, and many of the tools used to construct monitoring and data archival applications.

XML is not a good choice for moving I/O data around the factory floor. It would be highly wasteful to encode your 16 discrete valve states (on or off) in an XML document, and transfer each one as: "<valve_1>ON<\valve_1>" or "<valve_1>OFF<\valve_1>".

Summary

XML is a data encoding option for OPC UA. OPC UA Clients can receive transmissions from OPC UA Slaves where the data is encoded as XML. This option makes OPC UA attractive to many web service applications that already use XML to communicate with other IT applications.

XML is also used in technologies like MTConnect. This Device-Information Model specifies the organization and contents of a machine tooling system that is encoded in XML. The Streams Information Model that transfers values about a tooling system is also an XML file. The entire MTConnect technology is based on XML.

XML is more of a compatible technology with OPC UA than a competitive technology. Yes, it's true that you can accomplish many of the same information transfers using XML as a standalone technology, but you would miss the secure authentication and authorization, the secure messaging, the efficient binary transfer, and the scope of services; these provide a vastly greater number of capabilities than what is offered just by XML.

MTCONNECT

Those of us in the world of discrete and process automation think that everyone is part of the discrete world—of individual parts—or the process world—of continuous processes. The belief is that everyone uses PLCs and DCSs to control their systems, and that their network architectures are composed of the technologies most familiar to us: Modbus, Profibus, DeviceNet, EtherNet/IP, and ProfiNet IO.

That's a very limited view. There's a whole other world of automation devices – a pretty big one, at that – that uses very sophisticated automation systems to generate products. That "world" is the machining or automated tooling domain. That's a world with sophisticated cutting and machining tools, specialized processes, and its own terminology and methodologies.

MTConnect, which started in 2006, is the communication standard in the machine tool industry. Prior to MTConnect, shop floor machine monitoring of a series of disparate tools from different vendors was impractical for all but the largest, most profitable, and most technologically adept machine shops.

Overview

In Contrast to OPC UA, MTConnect is a pretty simple technology, one built on two very well-known standards: XML and HTTP. Very simply, MTConnect can be thought of as a well-defined standard for sending machine shop floor data as XML files. MTConnect uses simple HTTP Get instructions, the same instructions used for web Servers to deliver web pages, to request machining data from a controller.

This works very well in the tooling industry, as the industry is small enough to define a common vocabulary and the Information Model semantics to serve most, if not all, of the information transfer requirements in the industry. Items like Power on/off, Spindle Speed, Axis position, Axis Speed, Feed, Block number, Status, CNC mode, work number, Alarm, Spindle load, Axis load, Spindle Override, and the rest—all of these are common to most of the equipment in the industry.

MTConnect separates the software functionality needed to transfer data from a machine controller to an application into two components; Adapters and Agents. Adapters are software components that interact with the machine controller. An Adapter translates the machine controller data from its often proprietary values and formats it into the common terminology of MTConnect.

Agents are software devices that interrogate Adapters. Agents have two functions: 1) collect data from one or more Adapters and 2) provide that data to applications. Agents communicate with Adapters in any communication mechanism available from that particular Adapter. The MTConnect standard does not define how data moves from Adapter to Agent.

Agents are the software components that respond to HTTP Get requests from applications with XML files. Those requests can include Probe requests, which are requests for the XML Schema for a machine controller, or stream requests, which are data requests. The XPATH language is supported, which provides the capability to limit requests to specific values.

Summary

MTConnect is an open, royalty-free communications standard intended to foster greater interoperability between machining equipment and software applications on the shop floor. Those goals are similar but less encompassing than the goals of the OPC UA standard. OPC UA has the broader goal of providing highly secure, bidirectional interoperability among all types of manufacturing equipment, using a very flexible Information Model.

MTConnect is now a companion standard for OPC UA. What that has meant for other technologies is that the Object Model of the technology (XML Schema, in this case), can be transported using the secure and reliable channel/session infrastructure of OPC UA. That means more transports more connectivity, and more secure and reliable transfer of machining data into the enterprise.

HTTP

The Hypertext Transfer Protocol (HTTP) is the connectionless, stateless protocol that is used every time we access a web page. It is included here as more than a few vendors implement it as a very simple way of moving data between automation devices and IT applications.

What Is HTTP?

HTTP is a request/response protocol. A Client establishes a TCP connection with a Server and sends an HTTP request to the Server. The request generally includes a URL, the protocol version, and a message containing request parameters, Client information, and sometimes a message body. The Server responds with a status line which includes the protocol version, a response code, and a message body.

Unlike many other computer protocols, HTTP closes the connection when the request is complete. A new request means that another connection must be opened. This requirement to open a connection on each request and not carry over any information from a previous request is what makes HTTP a stateless protocol.

An HTTP message contains either a GET request to retrieve information from the remote system, a PUT (or POST) request to send information, or a HEAD request, which returns everything the GET request does except the message body.

In its most often-used implementation, the HTTP GET service is used to request a web page from a remote Server. The response message contains the Hypertext Markup Language (HTML) that forms the web page content that appears in your browser.

How Is HTTP Used?

HTTP is a very simple technology and many vendors have built applications on top of it to move data from the automation devices to IT applications. It is not difficult for vendors to customize HTTP GET and POST messages and add custom protocol information in the message body. By building applications that use these protocols, these vendors create easy to use mechanisms for moving automation data between IT systems and the factory floor.

For example, there is a device vendor that makes generic networking modules. This vendor supplies the hardware and software for building networked applications. Users write C code to build applications with the vendor's hardware.

By adding an API for cloud applications, the user can easily move data from this hardware to applications in the cloud that also use the vendor's API. Users, who are not concerned about using proprietary technology, can easily create cloud applications which manipulate and store data from the embedded hardware.

Summary

HTTP, like XML, is another component used by OPC UA to transfer data. Unlike its implementation as part of HTTP, when it is used in a standalone manner, there is no Information Model, no services other than the raw GET and PUT, and no standardized mechanism to move data. Most implementations based on HTTP use a proprietary body that encodes commands and responses.

OPC UA uses HTTP as a standard transport. HTTP together with

Simple Object Access Protocol (SOAP) forms one of the main transports used to link OPC UA devices with IT applications. SOAP provides the structure for embedding the OPC UA messages that open channels and sessions, configure subscriptions, read and write attribute values, and perform all the other sophisticated services in OPC UA.

REST

Unlike the other concepts described in this chapter, REpresentational State Transfer (REST) is not a protocol and not a technology but actually an architectural concept for moving data around the Internet. The REST architecture or a RESTful interface is simply a very flexible design, usually built on top of HTTP, for Client devices to make requests of Server devices using well-defined and simple processes.

What makes REST so significant is its widespread acceptance for many important applications as a simpler alternative to SOAP (Simple Object Access Protocol) and WSDL (Web Services Description Language). Leading companies like Yahoo, Google, and Facebook have passed beyond SOAP and WSDL-based interfaces in favor of this easier-to-use, resource-oriented model to expose their services.

Not many people are aware that every time they browse a web page they are using the REST architecture.

What Is REST?

In REST, the concept of how devices on a network function is different than the conceptual view of a network for most other networking technologies. We usually think of a network as a set of devices that provide some specific set of services. A Modbus device, for example, provides a specific set of services like Read Coil, Read Holding register, Write Coil, Write Register Single. EtherNet/IP Adapter devices and other CIP (Common Industrial Protocol) devices provide services like Read Attribute and Write Attribute. In most technologies used in Industrial Automation there is some set of predefined services that Client devices must learn, implement, and use to access the resources of a device.

This architecture works well in tightly controlled systems that serve smaller problem domains. For many years, Modbus was the accepted way to pull data out of energy meters. Read Register was the command to get values from energy meters. In well-defined paradigms like energy data collection, these kinds of architectures make sense.

In these systems it was just accepted that if a Server device like an energy meter implemented a new service like the capability to track detailed energy usage over a small period of time in great detail, all the Clients that wanted access to that new resource would be adapted with new software to use that feature.

That sort of architecture works well in these limited paradigm systems but it doesn't work well in the world of the World Wide Web. There, we have a case of unlimited, ever changing resources being made available in vast amounts. We don't want to update our Clients (web browsers) every time someone adds to a web page we visit every day.

It would be massively impractical to update our web browsers every time a change was made to a web page. The reason we don't do that is that the interaction between your web browser and the web Server is an example of a RESTful interface. Your web browser does an HTTP GET which retrieves markup text containing text and hyperlinks to other web pages. You now know more about the resources offered by that Server and can request more information by clicking on one of the hyperlinks. That click repeats the process, another HTTP Get is issued, a new set of markup text is delivered, and you have access to more pertinent resources.

This works because a different mindset is in place about Servers and the resources they offer. This mindset is resource-centric instead of function-centric. In the RESTful architecture, a Server is viewed as a set of resources, nouns if you will, that can be operated on by a simple set of verbs like GET, POST, UPDATE and the like. This architecture yields a much more flexible mechanism for retrieving resources than the limited function-centric kinds of technologies we've used in the past.

There are just a few principles that define REST:

- REST is an architecture style – all web interactions can be said to be REST architecture operations.
- REST is not a standard – there is no specification for it from some standards body.
- REST is a stateless Client-Server protocol. Stateless meaning that the Server remembers nothing from any previous interaction with that Client.
- REST is based on HTTP but there is no reason it can't be implemented on any other transport.
- RESTful applications use the simple HTTP services called CRUD (Create/Read/Update/Delete).
- Like other Web Services, REST has no built-in security, it offers no encryption, doesn't do session management or any other added-value service. These can be made available by adding components to the transport or by using a different transport like HTTPS.

How Is REST Used?

Yahoo and Facebook have created RESTful Client APIs which simplify the process of accessing resources within their systems. These APIs have largely replaced the remote procedure call technologies previously used, like SOAP and WSDL. In fact, it's clear that all these types of remote procedure call technologies are obsolescing and taking their place in Internet history.

But, you might say, it's clear that the REST architecture is perfect for browsers and humans accessing web pages but how is it used for machine to machine (M2M) kinds of communication?

That's one of the reasons that HTTP is so perfect for a REST architecture. REST has a built-in feature that allows the Client to select the format in which the Server should return the resource. Web browser Clients use the resource to return markup language (HTML) that can be displayed. Machines, on the other hand, can request things like Java objects, XML, CSV files, and other data more easily processed by machines.

Summary

REST is a very good alternative to OPC UA. It is simpler, easier to implement, but less functional. As a simple mechanism to move factory floor data to an IT application or cloud Server, REST can be a good choice. You can implement a factory floor Server that provides a REST interface, and define Java objects, XML, or CSV as the delivery format for your data. It won't be real time – but you don't always need real time data.

With OPC UA you have some capabilities you won't get with a REST interface. You have Publish/Subscribe capabilities so that you can get data pushed out on schedule. You have the capabilities to move binary data, and you have access to sophisticated Information Models that are used in your industry.

In general, OPC UA is a more formal mechanism for providing very secure, reliable, structured information from the factory floor to IT applications and the cloud. REST is a more informal mechanism. The choice is really application dependent.

MQTT

Overview

MQTT (Message Queuing Telemetry Transport) is another mechanism for moving data around the factory floor or from the manufacturing environment to the cloud. MQTT is designed to meet the challenge of publishing small pieces of data in volume to lots of consumers' devices constrained by low-bandwidth, high-latency, or unreliable networks.

MQTT supports dynamic communication environments where large volumes of data and events need to be made available to tens of thousands of Servers and other consumers.

What Is MQTT?

The heart and soul of MQTT is its Publish/Subscribe architecture. This architecture allows a message to be published once and go to multiple consumers, with complete message decoupling between the producer of the data/events and the consumer(s) of the messages and events.

Data is organized by topic in a hierarchy with as many levels of subtopics as needed. Consumers can subscribe to a topic or a subtopic. They can also use wildcards to specify topics of interest.

Namespace designations are used to identify topics. A broker receives the information from the Servers and matches the information to consumers by topic subscriptions. The information is distributed to every consumer that matched the topic. If no consumer requires the information, the information is discarded.

Topics are designated by a namespace. Subtopics are designated with a slash ("/"). An energy system might publish information about itself on:

<HouseID>/<system>/<meternum>/energyConsumption

Where:
HouseID identifies a specific location.
System Identifies the HVAC, Kitchen, or solar system.
Meter Number identifies a specific meter in a system.

Consumers can subscribe to all meters in a system with HouseID/Kitchen/* or all systems with HouseID/*.

What Are Benefits of MQTT?

Some of the benefits of MQTT technology include:

- Very efficient event handling protocol. MQTT is a "PUSH" system in which the producers push data to brokers. No bandwidth is consumed by consumers requesting data.
- Low latency. Information is pushed immediately to consumers.
- Low resources required by publishers. This makes it very good for low-resourced devices like sensors and actuators.
- Very reliable operation on fragile and unreliable networks. Brokers can be configured to retain messages for consumers that are temporarily disconnected.

Summary

MQTT is a very simple way of distributing information from lots of publishers to lots of consumers. It is extremely lightweight, reliable, and adapts well to low resourced devices. Broker devices, which some view as a disadvantage, manage the connection between the publishers and consumers.

OPC UA is much larger, more complete, and not well adapted to low resource devices. MQTT is a superior way of moving data from sensors and actuators to various kinds of consumers. The Publish/Subscribe technology of MQTT is vastly superior to OPC UA, as much more bandwidth is consumed by OPC UA Publish/Subscribe operations.

OPC UA advantages over MQTT include the superior Information Modeling, built-in security, and more robust Service Oriented Architecture. OPC UA Clients have more capabilities for managing and accessing Server data than MQTT.

DDS

Overview

The Data Distribution Service (DDS™) is an architecture for data-centric connectivity. DDS provides more complete communication services than other IIoT (Industrial Internet of Things) communication systems. It provides automatic data discovery, security, content scheduling, and other management tasks that fall to applications in other technologies.

DDS is designed to manage a large network of publishers and subscribers by focusing on the data being communicated instead of the communication method. Instead of messaging, DDS focuses on the contextual relationships of the data, and organizes publishers and subscribers around what data they need and when they need it.

Unlike many other IIoT protocols, DDS completely decouples subscribers from publishers. Geographic location, redundancy concerns, time sensitivity, data distribution, and platforms are all decoupled by the DDS architecture.

What Is DDS?

The heart and soul of DDS is the concept of data centricity. Data appears to each application simply as a local store. An application interacts with that local data store using its native data model. A Java application, for example, works with Java objects. The application uses the data as if it is not part of the IIoT. DDS manages that local data store and connects it to other publishers and subscribers of data on the network without specific guidance from the application.

Data is organized by topic within a domain. Domains are a logical scope for a set of topics. A data writer and a data reader can only match topics if they join the same domain. Topics can be shared or exclusively owned. There can be multiple topics and instances of topics. Topics can be filtered as needed so information flows only when it is relevant and necessary.

Like MQTT, OPC UA, and other technologies, DDS uses a Publish/Subscribe communications model. Data writer applications write data into their local data store that DDS makes available to data readers who subscribe to topics available in the domain.

DDS does this automatically by matching data writers and data readers.

Consumers, or subscribers, interested in a topic can declare an interest in that topic. DDS matches that interest to a publisher or data writer offering that data. The matching of writers to readers is entirely managed by DDS and not the application code. It is an anonymous arrangement. The data writer has no information how many data readers, if any, are interested in its data.

There are two architectural elements to DDS: Data-Centric Publish/Subscribe (DCPS) and Data Local Reconstruction Layer (DLRL). DCPS is the portion of DDS that manages the exchange of topic data with other DDS enabled systems. DLRL is the portion of DDS that manages access to topic data from the local application.

What Are the Benefits of DDS?

Some of the benefits of the DDS technology are reported to be:

- DDS can handle timing-critical distributed systems better than any other IIoT competitor.
- Applications using DDS are decoupled from traditional network interactions like determining who should receive messages, when they should receive them, and what to do if something goes wrong.
- DDS automatically matches writer to reader without guidance from the application.
- Applications only interact with what they perceive as a local data store.
- Applications interact with the local data store in their native data format.
- DDS is very tolerant of network delays.
- DDS offers low latency. Information is immediately pushed to consumers.
- Content can be filtered by the DDS network to eliminate data that is not needed by a subscriber.
- DDS delivers very reliable operation on fragile and unreliable networks.

Summary

It is not really fair to compare OPC UA and DDS; each meets a different design goal. OPC UA provides an architecture for moving information from devices to IT applications. DDS is a device-to-device architecture. It is an IoT protocol that simultaneously moves millions of messages per second to numerous receivers. Though DDS is capable of moving data to IT applications, it isn't designed for that.

DDS is very impressive technology. Its Publish/Subscribe architecture appears to be much better than that offered by OPC UA because of its device-to-device design. The ability to easily integrate a large number of nodes into a domain and easily share topic data is remarkable.

As of this writing the OPC UA Publish/Subscribe architecture is being redesigned, but it will never compete with the performance of DDS. DDS is certainly going to be better at delivering time sensitive data than OPC UA ever will. That is not a design goal of OPC UA. The TSN (Time Sensitive

Network) architecture is being integrated into OPC UA, but how well that is going to work is an unknown.

Both DDS and OPC UA have their place. OPC UA is clearly better at Information Modeling and interactions with IT applications. DDS is better at high performance device-to-device messaging.

TERMS TO KNOW

One of the things that you'll learn about OPC UA is that the terminology is a little different than what you'll be used to seeing. No matter how many years of industrial networking experience you have, you may be unexpectedly confused when you start to study OPC UA.

A lot of the terms are similar to what you might expect, but the meanings are twisted slightly. Because OPC UA is a technology that really crosses the line between the enterprise and the factory floor, the terms can be confusing to well-versed individuals in both the IT and factory floor worlds.

Another reason why the terms appear at first glance to be a little confusing is the scope of OPC UA. In IA (Industrial Automation), we generally talk about interfaces between software components cohabiting in a processor. Or we talk about devices on the same subnet communicating over a very well-defined and very restrictive interface (EtherNet/IP, Modbus TCP or ProfiNet IO). In the OPC UA world, we talk much more about generic services with much more flexibility and capability than the interfaces between factory floor devices.

Here are a number of those common OPC UA terms that you'll need to know:

Application Layer – In industrial networking, this term refers exclusively to the end user application and not the protocol stack, whatever it may be. The end user application implements some set of defined functionality and moves well-defined data between the application and some external device using a very restrictive interface. This isn't quite the same in OPC UA. In OPC UA, the application usually encompasses the end user application, the object model, and the set of OPC UA services implemented by the device. This is a much more all-embracing use of the term.

Application Program Interface (API) – An API is the set of software interfaces that allow one software application to use the services of another software application. In the industrial world, this normally refers to the interface between two pieces of software cohabitating in the same processor. In the Ethernet world, the API can refer to the interfaces needed by some Client device to access the available services of some remote web service. In OPC UA, the API generally refers to the set of interfaces that an OPC UA toolkit vendor provides to a device developer. Because different toolkits are designed differently, the APIs work differently. The API may include interfaces to the data model. In other cases, the API may only interface the three main components of OPC UA: the encoding layer, the security layer, and the transport layer. The OPC UA data and service model may be part of the user application.

Asymmetric Security – In Asymmetric security, both sender and receiver have a private key that they keep secret and a public key that they share with anyone. Messages to the key holder are encrypted with the public key and decoded by the key holder with the private key. Message segments that require verification of origin are signed with the private key and validated using the public key.

Auditing – The recording of all actions, activities, users, resources, and more in a system. In OPC UA, auditing is incorporated as a normative part of the specification, as a mechanism for the system administrators to identify vulnerabilities and diagnose security breaches.

Authentication – Authentication means to verify who the sender is. Usually that means to validate the sender's X509 certificate and verify that the certificate is currently valid and was signed by a reputable and trusted Certificate Authority.

Authorization – Authorization is the process of validating access to a resource. Once a sender is authenticated, the sender must be granted access to resources in the receiver. Authorization can be accomplished using a list of trusted names, a user name and password, or any other reliable mechanism.

Binary Encoding – OPC UA Binary Encoding is a way to serialize data using an IEEE binary encoding standard. An encoding is a specific way of mapping a data type to the actual data that appears on the wire. In binary encoding, data is mapped to a very compact binary data representation that uses fewer bytes and is more efficient to transfer and process by embedded systems. Binary encoding is widely used by Industrial Automation systems, but less common among enterprise applications.

BLOB (Binary Large Object Block) – "BLOBs" provide a way to transfer data that has no OPC UA data definition. Normally, all OPC UA data is referenced by some sort of data definition that describes the structure and typing for the entity. BLOB data is used when the application wishes to transfer data that has no OPC UA definition. BLOB data is user defined and could be anything: video, audio, data files, or anything else.

Certification Authority (CA) – An organization that creates and distributes Digital Certificates. The CA creates the public and private keys that are associated with the certificate owner. The CA often encrypts a portion of the certificate with its private key (signs it) to provide assurance to anyone in the future that the CA did create the certificate. Of course, sending a certificate to a receiver is only effective if the CA is known by the receiver to be an honest and reputable certificate provider.

Client – An OPC UA Client is the side of OPC UA communication that initiates a communication session. Clients in OPC UA are much more flexible than other network Clients. OPC UA Clients have the capability to search out and discover OPC UA Servers, discover how to communicate with the Server, discover what capabilities the OPC UA Servers have and configure the OPC UA Server to deliver specific pieces of data when and how they want it. OPC UA Clients support many different encodings, security mappings and transports so that they can communicate with all different types of Servers.

Digital Certificate – A sequence of data bytes that functions like your driver's license. The Digital Certificate verifies that you are who you say you are. There are many components to a Digital Certificate, including the name of the algorithm, the organization that created it, the public key of the owner, and the dates it is valid. X.509 refers to the most popular certificate standard. You will often encounter the term "X509 Certificates." You will also encounter the term "DER Certificates," which refers to a method for encoding certificates as a series of bytes.

Digital Signature – Also called signing. A digital signature is a small series of bytes that are the result of processing some larger series of bytes through an algorithm. The resulting smaller series of bytes are encrypted with the owner's private key. The receiver can verify that the true owner "signed" the document by decoding the encrypted result using the owner's public key and validating the result using the same algorithm the sender used. Signing a document or message guarantees the integrity of the portion of the message signed with the owner's private key.

Data Encoding – A data encoding is a specific a way of converting an OPC request or response into a stream of bytes for transmission. Two encodings are currently supported in OPC UA: OPC UA Binary and XML. OPC UA Binary is a much more compact encoding with smaller messages, less buffer space and better performance. XML is a more generic encoding that is used in many enterprise systems. XML is easier for enterprise Servers to process, but requires more processing power, larger messages and more buffer space.

HTTP (Hypertext Transfer Protocol) – is part of the basic plumbing of the Internet. It is the low-level protocol that allows a Client application like your browser to request a web page from a web Server. HTTP messages request data or send data in a very standard format supported by every Internet-aware application.

HTTP/SOAP Transport – The HTTP/SOAP transport is a transport currently supported in OPC UA. This transport requires larger messages, bigger buffers and more processing, but is used because HTTP and SOAP are supported by almost all (if not all) enterprise applications. It is a standard way of moving serialized OPC UA messages between a Client and a Server.

Mappings – This is an interesting term. The OPC UA specifications are very abstract, unlike, say, a Modbus RTU specification. Modbus RTU runs over multi-drop RS485 and that fact is inherent in the specifications. It's not that way with OPC UA. The specifications for OPC UA operation are very abstract and done that way to maintain the ability to take advantage of future technologies. A mapping refers to how those abstract specifications are mapped onto a specific technology. For example, an encoding mapping describes how a generic OPC UA message is serialized. A security mapping describes how security is applied to that encoded message. A transport mapping describes how that message is structured for transmission over a TCP connection.

NodeID – The the identifier for everything in the Server address space. It classifies every node and provides the means for a Client to identify the origin of a node. Knowing the naming authority that created the node means that the Client can identify that two nodes in two different Servers refer to the same node, i.e. a speed variable in a motor drive is the same as a speed variable in another motor drive.

Private Key – A private key is a key that an owner keeps private and never releases to anyone else.

Protocol Stack or Stack – A protocol stack in industrial networking, like EtherNet/IP or ProfiNet IO, generally implements the data model and the services of that protocol. An API connects that data model and service model to the data of the end user application. Though protocol stack vendors can implement this in many different ways, in general when the OPC UA protocol stack it discussed, the reference is only to the core of OPC UA communication. That doesn't include the data encoding, security, or network transport. It also doesn't include the data representation, Service Set, or address space. In OPC UA the protocol stack represents just a small part of OPC UA.

Public Key – A public key is a key that an owner makes available to everyone who requests it.

Public Key Encryption – A cryptographic system that uses two keys. One key is kept private. The other is made public. Private messages can be exchanged in either direction using the two keys.

PKI (Public Key Infrastructure) – Technically, this is a set of hardware, software and policies needed to manage certificates, keys, access lists, and the keys used in Public Key Encryption. In our world (manufacturing automation), PKI simply refers to a system in which every device has two keys, a public key and a private key.

RSA – RSA is a very popular public key cryptography algorithm. RSA refers to the initials of the three designers of the algorithm. RSA with various size key lengths is used in OPC UA and other popular and secure protocols.

Security Mapping – A security mapping is the mechanism that ensures the integrity and privacy of messages being transferred across a connection. In OPC UA, the security mapping is applied to message chunks, pieces of a message that are created to efficiently use the bandwidth of the communication media. The security mapping is independent of the protocol or byte stream encoding. Several different types of security mappings are available in OPC UA.

Security Protocol – A security protocol protects the privacy and integrity of messages. OPC UA takes advantage of several standard, well-known security protocols. The selected security protocol for a specific application is a combination of the security requirements for the installation and the encoding and transports selected for OPC UA implementation.

Serialization – The process of transforming a message like the OPC UA Read Attribute service into a series of bytes that can be understood by another device. Serialization dictates how data elements like a floating point value are transformed into a series of bytes that can be sent serially over a wire. Two types of serialization, or encodings, are currently supported by OPC UA: OPC UA Binary and OPC UA XML.

Server – An OPC UA Server is the side of OPC UA communication that responds to service requests from an OPC UA Client device. There is no standard OPC UA Server either in functionality, performance, or device type. Devices ranging from small sensors to massive chillers may be OPC UA Servers. Some Servers may host just a couple of data points. Others might have thousands. Some OPC UA Servers may use mappings with high security and lower bandwidth XML encoding, while others may communicate without security using high performance OPC UA Binary Encoding. Some Servers may be completely configurable and offer the Client the option to configure data model views, alarms, and events. Others may be completely fixed.

SHA – SHA is a series of public key cryptography algorithms published by the National Institute of Standards and Technology (NIST). SHA algorithms are also used in OPC UA to sign and encrypt messages.

SOAP (Simple Object Access Protocol) – SOAP extends XML and provides a higher level of functionality. Among other things, SOAP adds the ability to make remote procedure calls within an XML structure.

Symmetric Security – In Symmetric security, both the receiver and the sender hold the key to decrypt messages. One encrypts with the symmetric key, the other decrypts with it. It is called symmetric since both hold the same key.

Transport – A transport is the mechanism that moves an OPC UA message between a Client and Server. This is another term that at first glance can be confusing. All OPC UA messages are delivered over a TCP/IP connection. TCP/IP provides applications with reliable channels of communication across a network over an acknowledged connection. A transport layer, or in the terms of OPC UA, a transport mapping, is a mechanism for providing additional transport layer services. These include ordering of messages, and separating messages into chunks that can be accommodated by the other device given its buffer and message size limitations.

Transport Protocol – A transport protocol is the technology that actually provides the end-to-end transfer of OPC UA messages between OPC UA Clients and Servers. Once an OPC UA service message is encoded and passes through securitization, it is ready for transport. Two transports protocols are currently defined for OPC UA: OPC UA TCP and SOAP/HTTP. The underlying technology for both these transports is standard TCP. TCP provides the socket-level communication between Clients and Servers.

UA TCP Transport – UA TCP transport is essentially a small protocol that establishes a low-level communication channel between a Client and a Server. Most of what the UA TCP transport does is to negotiate maximum buffer sizes so both sides understand the limits of the other. The advantage of UA TCP is its small size and negligible impact on throughput.

Web Services – Web Services is a generic term for loosely coupling Internet services (applications) in a structured way. The majority of Internet applications today are built using Web Services. With Web Services, you can easily find services, obtain the interfaces and characteristics of the interfaces and then bind to them. HTTP, SOAP, and XML are the basic technologies of Web Services applications and are some of the technologies that can be used by OPC UA Clients and Servers.

XML (eXtensible Markup Language) – XML is a highly structured way of specifying data such that applications can easily communicate. XML transfers all data as ASCII – the one commonly understood data format for all computer systems. XML uses a grammar to define the specific data tags that are used by an application to pass data.

XML Encoding – XML encoding is a way to serialize (encode) data using eXtensible Markup Language (XML). An encoding is a specific way of mapping a data type to the actual data that appears on the wire. In XML encoding, data is mapped to the highly structured ASCII character representation used by XML. XML can be cumbersome, large, and inhibit performance, but the encoding is used because a large number of enterprise application programs support XML by default.

RESOURCES

There are two excellent books on OPC UA that are much more in-depth than this this one. These are a good follow-on to this introduction.

Juergen Lange, Frank Iwanitz and Thomas J. Burke. <u>OPC from Data Access to Unified Architecture</u>– From Data Access to Unified Architecture. VDE Verlag, Berlin, Germany 2010

Wolfgang Mahnke, Stefan-Helmut Leitner and Matthias Damm. <u>OPC Unified Architecture Textbook</u> – The fundamentals and theory behind OPC Unified Architecture. Springer-Verlag, Berlin, Germany 2009

ABOUT THE AUTHOR

John Rinaldi, is Chief Strategist, Business Development Manager and CEO of Real Time Automation (RTA) in Pewaukee, WI.

After escaping from Marquette University with a degree in Electrical Engineering (graduating cum laude, no less), John worked in various jobs in the Automation Industry before once again fleeing back into the comfortable halls of academia. At the University of Connecticut, he once again talked his way into a degree, this time in Computer Science (MS CS).

John achieved marginal success as a Control Engineer, a Software Developer and IT Manager before founding Real Time Automation because "long term employment prospects are somewhat bleak for loose cannons."

With a strong desire to avoid work, responsibility and decision making, John had to build a great team at Real Time Automation, and he did. RTA now supplies network converters for Industrial and Building Automation applications all over the world. With a focus on simplicity, US support, fast service, expert consulting, and tailoring for specific customer applications, RTA has become a leading supplier of gateways worldwide.

John freely admits that the success of RTA is solely attributed to the incredible staff that like working for an odd, quirky company with a single focus: "Create solutions so simple to use that the hardest part of their integration is opening the box."

John is a recognized expert in industrial networks and the author of six books, four on industrial networking. The Industrial Ethernet Book focuses on explaining Industrial Ethernet concepts in a straightforward, clear fashion. The same simplicity is found in John's other book, OPC UA: The

Basics, an overview of the enhancements to OPC technology that allow for enterprise communication. A book on Modbus Technology, Modbus: The Everyman's Guide to Modbus, is popular with engineers new to Industrial Automation. All John's books are available on Amazon.

You can reach John here:

John Rinaldi
Real Time Automation
N26 W23315 Paul Rd
Pewaukee, WI 53072
262-436-9299 (office)
414-460-6556 (Cell)

http://www.rtaautomation.com/contact-us/
https://www.linkedin.com/in/johnsrinaldi

LEARN MORE Have a little fun and get some relevant information. Sign up for the Real Time Automation Newsletter

http://www.rtaautomation.com/company/newsletter/

Made in the USA
San Bernardino, CA
17 September 2016